STRUCTURE AND FUNCTION OF CELLS

To Hilary, Sally and Laurence

STRUCTURE AND FUNCTION OF CELLS

*A text for students in medicine
and science*

COLIN R HOPKINS BSc PhD

*Professor, Department of Histology and
Cell Biology, The Medical School,
Liverpool University*

1978

W. B. Saunders Company Ltd   London · Philadelphia · Toronto

W. B. Saunders Company Ltd: 1 St Anne's Road
Eastbourne, East Sussex BN21 3UN

West Washington Square,
Philadelphia, PA 19105

1 Goldthorne Avenue,
Toronto, Ontario M8Z 5T9

Library of Congress Cataloging in Publication Data

Hopkins, Colin R.
    Structure and Function of Cells

    1. Cytology.   I. Title.
QH581.2.H66   574.8'7   77–91850

ISBN 0–7216–4775–8

Printed and bound in Great Britain by Butler & Tanner Ltd, Frome and London

Print Number: 9   8   7   6   5   4

PREFACE

Over the last decade the rapid, almost explosive growth in our understanding of structure-related function has served to emphasize the central role that cell biology has to play in the education of students in medicine and science. With its recent progress, cell biology has now developed sufficiently to provide a focus in which the surrounding, related areas of interest in biology can be integrated with those in which attention is directed towards the molecular level. The biology of the cell is thus beginning to be taught as a unifying topic in its own right rather than as an appendage to courses in morphology or biochemistry. The purpose of this text is to convey this view to students taking preliminary courses in basic medical science.

The book is intended primarily for students who have a background of school biology and who are taking concurrent courses in biochemistry and physiology. As a brief, contemporary assessment of the field, it should also provide a useful basis for later courses in pharmacology, histopathology, immunology and experimental medicine in general. However, although the approach is clearly directed at the eukaryotic system as typified by the cells of mammalian tissues, it will be read without difficulty by students in life science who lack a strong medical bias. It is with these students in mind that a glossary has been included.

As this book may be the only exposure a student has to the topic of cell biology, I have, in writing it, tried to make the text accessible to first-year university students whilst maintaining something of the momentum and excitement that pervades the subject at the present time. For this reason the more established aspects have often been treated in a condensed and abbreviated manner and the bibliography has been used to provide some indication of historical perspective. The chief dangers of this approach are of course that the generalizations will be too sweeping and the promise of the more recent areas of advance will in time be shown to have been evanescent. Being aware of a danger helps, of course, but is no guarantee that it can always be avoided.

ACKNOWLEDGEMENTS

In my search for suitable illustrative material I have received generous help from workers in many different laboratories. To all of them I am most grateful. For much of the artwork I have been very fortunate in being able to rely upon the talent and patience of Ken Biggs. I am grateful also to Nancy K. Dwyer.

Some of the electron micrographs have been produced in my laboratory and for them I have depended upon the expert technical assistance of Hazel Smith. For the background, often tedious, jobs that have to be done in compiling a text of this kind I am pleased to acknowledge the willing and often innovative help of Annie Pritchard. For much of the typing I am grateful to my secretary, Mrs Doreen Godsell.

To my departmental colleagues and students who over the years have both knowingly and unknowingly made their contributions, I wish to express my thanks.

Finally I thank my wife for her loving support and constant encouragement.

Colin R. Hopkins

CONTENTS

LIST OF ILLUSTRATIONS

PART 1

Techniques in the study of cell structure

Much of our current understanding of cellular structure and function is very closely bound up with a proper appreciation of the strengths and weaknesses of the available technical methods employed in their study. It will, therefore, be instructive for us to preface our discussion of the cell with an account of the most widely used microscopes and their associated preparative and applied techniques.

THE MICROSCOPE

MAGNIFICATION AND RESOLUTION

In their most basic essentials all microscopes aim: (a) to magnify the object, and (b) to display the object in greater detail. These aims are interdependent and it is important to realize that to increase magnification without a commensurate improvement in the degree of discernible detail is of little advantage.

Magnification

The magnification of any optical system is dependent upon the focal length of the lenses in the system and their mutual arrangement. It is usually expressed as the ratio of the length of the final image to that of the object, and for the ordinary class microscope it is usually between ×25 and ×1500.

Resolution

The resolving power of a lens indicates the fineness of detail that it allows to be seen. Thus, if one examines two small objects with a microscope lens, provided the objects are well separated, they will be resolved as separate entities. If, however, they are then gradually moved closer together, a situation will eventually arise in which the two objects, though still separate, can no longer be seen to be distinct from each other. In this situation, only by improving the resolution (i.e. by using a lens with better resolving power) will it again be possible to render the two objects as separate entities (see Figure 1).

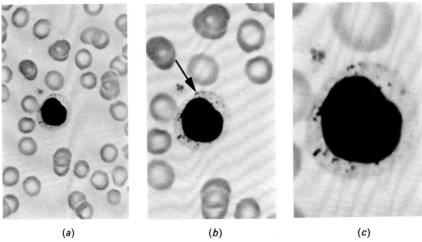

(a)   (b)   (c)

Figure 1. *Resolution and magnification. A leucocyte in a blood smear photographed at increasing magnifications with the same objective lens (×100; NA=1.32). Thus, although in Figures (a) to (c) the final magnification is increased, the resolving power remains the same.*

In (b), at a magnification of ×1150, two granules (arrowed) within the leucocyte are resolved and seen as separate entities, while in (a), at a lower magnification of ×650, although the lens resolves the granules, they are not seen to be separate. In (c), where the magnification is increased to ×2250, the granules are larger than in (b) but there is no more detail to be seen.

Increasing the magnification over (b) without increasing the resolution is thus of no advantage (i.e. it produces 'empty magnification').

Courtesy of J. James, Histologisch Laboratorium Amsterdam, University of Amsterdam.

Factors that determine resolution

1. Numerical aperture

When light rays pass through a specimen containing fine detail they interfere with each other and they are variously diffracted; increasingly fine detail increases their angles of diffraction. Since the resolving power of a lens depends upon its ability to collect these diffracted rays, the wider the angle of rays collected the better is the resolution.

The capacity of a lens to collect rays emerging from an object is defined by its *numerical aperture* (NA), and this depends upon both its *angular aperture* (u in Figure 2) and the *refractive index* (n) of the medium through which the rays pass.

The relationship is expressed as:

$$NA = n \cdot \sin u$$

In any given lens, the NA (and thus the resolution) is at its best when the cone of rays emerging from the object just fills the angular aperture. When setting up a microscope this optimum requirement is only obtained by careful focusing of the illumination system (see section below relating to the substage condenser).

In the conventional light microscope the medium between the low-power (less than ×40) objective lenses and the specimen is air (i.e.

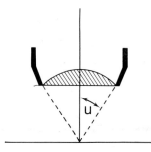

Figure 2. *The angular aperture of an objective lens.*

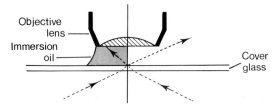

Objective
lens
Immersion
oil
Cover
glass

Figure 3. *Oil immersion. Light rays emerging from the coverglass into air (i.e. from a dense to a less dense medium) will be bent by refraction towards the glass. Immersion oils (such as cedar wood oil) have the same refractive index as glass and thus refraction is much reduced.*

the refractive index, $n = 1$). However, for lenses of higher power, where maximum resolution is required, the refractive index may be increased by filling this space with a special 'immersion' oil. The refractive index of the immersion oils used with glass-covered microscope slides is optimally about 1.55. As indicated in the equation above and as shown in Figure 3, this arrangement increases the NA and results in fewer light rays being lost due to refraction. Resolution is thus improved.

2. Wavelength

Resolution also depends upon the wavelength of the transmitted wave form; the smaller the wavelength the better is the resolution. It is primarily for this reason that the electromagnetic lenses of the electron microscope, which depend upon the extremely short wavelength of the electron (0.005 nm at a 60 kV accelerating voltage), can resolve details that are orders of magnitude smaller than those resolved by the light microscope (see below).

The resolution limit (r)

The resolution limit of a lens (which is the converse of its resolving power) can be defined as the minimum separation between two points at which they can still be observed as two distinct entities. With reference to the numerical aperture (NA) and wavelength (λ), it is expressed as:

$$r = \frac{0.61\ \lambda}{NA}$$

1 millimetre
(1 mm) = 1000
micrometres
1 micrometre
(1 μm) = 1000
nanometres (nm)

In practice, the maximum numerical aperture available for light microscope objective lenses is about 1.4. The component of white light to which the eye is most sensitive is green light (wavelength 560 nm); therefore this equation indicates that the best resolution limit obtainable when using white light is about 240 nm (or 0.24 μm). (For reference: it is useful to remember that a red blood cell has a diameter of about 7 μm.)

The numerical aperture and general performance of electromagnetic lenses is, by comparison with their light microscope counterparts, relatively poor (their NA is about 0.008) and for this reason the potential increase in the resolution limit derived from the short wavelength of the electron is much reduced. In practice it is about 0.25 nm. (For reference: the plasma membrane of the cell is about 7.5 nm thick.)

THE BASIC PLAN OF THE LIGHT MICROSCOPE

In the conventional light microscope the lens systems are arranged to provide: (a) two stages of magnification, and (b) an adequate and controlled illumination.

Magnification – the image-forming lenses

As shown in Figure 4, the two-stage magnification is accomplished by the objective lens forming a real, enlarged image (within the barrel of the microscope), which, in turn, provides the object for the ocular (eyepiece) lens. The ocular provides the final (virtual) image focused on the retina of the observer's eye. The purpose of the ocular lens is thus to enlarge the image produced by the objective lens and allow the observer to discern the detail contained within it. This lens cannot improve upon the resolution provided by the objective lens.

The illumination system (refer to Figure 5)

The illumination system of a microscope is designed to provide an adequate and even light intensity. As shown in Figure 5, it usually consists of a light source (the filament of an electric lamp) and a series of 'field' lenses and diaphragms (or 'stops') arranged between the components of the imaging system. Field lenses are placed close to the image

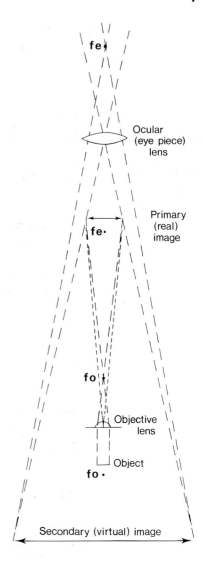

Figure 4. *The arrangement of lenses and two-stage magnifications in the conventional light microscope. fo — focal length of objective lens; fe — focal length of ocular lens.*

positions of the image-forming lenses so that they focus the light source near the image-forming lenses themselves. In this way the image of the source is out of focus in the focal plane of the image-forming lens and even illumination is provided.

Field diaphragms are positioned close to the image plane and, if they are adjustable, they can be focused along with the image. Their purpose is to eliminate stray light by controlling the size of the illuminated field.

Aperture diaphragms restrict the angles made with the optical axis by the image-forming rays and thus they serve to reduce glare. They may be fixed (in the rear focal plane of the objective, for example) or, as in the case of the substage condenser diaphragm, they may be adjustable.

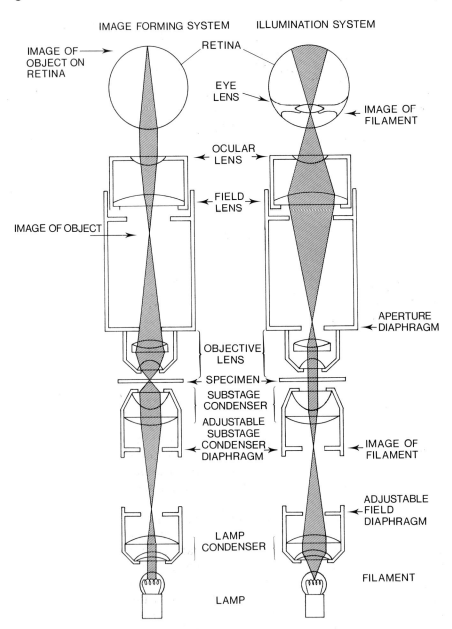

IMAGE FORMING SYSTEM   ILLUMINATION SYSTEM

IMAGE OF OBJECT ON RETINA

RETINA

EYE LENS

IMAGE OF FILAMENT

OCULAR LENS

FIELD LENS

IMAGE OF OBJECT

APERTURE DIAPHRAGM

OBJECTIVE LENS

SPECIMEN

SUBSTAGE CONDENSER

ADJUSTABLE SUBSTAGE CONDENSER DIAPHRAGM

IMAGE OF FILAMENT

ADJUSTABLE FIELD DIAPHRAGM

LAMP CONDENSER

FILAMENT

LAMP

Figure 5. *Imaging and illumination systems in the light microscope. The illmination system used in this example is the Köhler illumination system. It depends upon the condenser lenses immediately in front of the lamp focusing the image of the source in the rear focal plane of the substage condenser (i.e. at the level of the substage condenser diaphragm). This arrangement provides a homogeneous light source, which, as shown in the diagram of the image-forming system, is focused at the plane of the specimen.*

A simpler arrangement – the Nelson system – is still often used in classroom microscopes (see appendix at the end of this chapter). In this system the light source (usually an opal bulb) is imaged, via a plane mirror, directly on the object. The illumination obtained in this way is more uneven. After L. H. Greenberg.

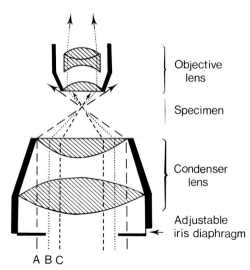

Objective lens

Specimen

Condenser lens

Adjustable iris diaphragm

A B C

Figure 6. *The substage condenser system. The adjustable iris diaphragm controls the angle of the cone of light available to the objective lens. The setting shown (B) fully illuminates the angular aperture of the objective lens, but (as discussed in the text) to eliminate glare, this aperture is normally reduced even further (C).*

The substage condenser

The primary purpose of the substage condenser lens is to enable the objective to achieve its maximum NA (and thus maximum resolution) by providing an angular cone of light that just fills its angular aperture. The NA of the condenser lens must therefore match closely the NA of the objective.

Below the lens of the substage condenser there is an adjustable diaphragm that, as shown in Figure 6, limits the angle of the cone of light emerging from the condenser. For each objective lens the adjustment of this aperture is crucial, because it requires a balanced compromise between optimum resolution and optimum illumination. Thus if the aperture of this diaphragm is not opened enough, the full NA of the objective lens will not be used and its maximum resolution will not be obtained. If, on the other hand, the aperture is opened to the full extent required to fill the angular aperture of the objective (and provide its maximum NA), glare arising from imperfections in the lens will be introduced. Setting this diaphragm to allow the condenser lens to illuminate two-thirds of the objective lens is the compromise most usually adopted (see the appendix at the end of this chapter).

Clearly the iris diaphragm of the substage condenser should never be used alone to control the light intensity (i.e. as an aperture diaphragm). For this purpose the brightness of the lamp filament must be altered – preferably by neutral density filters.

THE BASIC PLAN OF THE ELECTRON MICROSCOPE

Under appropriate conditions, electrons (the particulate constituents of atoms) display wave properties. This means they have an associated wavelength and can, for example, be diffracted like other forms of wave motion such as visible light. Because of this, and because electromagnetic lenses can be designed to focus electrons in much the same way as converging glass lenses focus light rays, transmission electron microscopes are designed with the same basic plan as the conventional light microscope. (For convenience, the source and the lens arrangement in the electron microscope are usually inverted, see Figure 9.)

The electron gun

An electron gun provides the illumination source. As shown in Figure 7, it consists of a V-shaped filament (the cathode) together with a cathode shield and anode. Both the shield and the anode are arranged so that their central apertures are in line with the tip of the filament. The filament is made of tungsten and can be heated to incandescence for prolonged periods by passing a current through it.

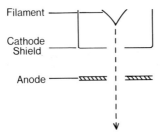

Filament

Cathode
Shield

Anode

Figure 7. *The arrangement of components in the electron gun and the path of the emerging electrons.*

Like other metals, tungsten, when heated, displays the property of thermionic emission and emits electrons. When they emerge from the tip of the heated filament, these emitted electrons are brought under the influence of a large potential difference maintained between the negative filament and cathode shield and the anode by a high accelerating voltage (40 to 100 000 volts). The gun thus produces a narrow beam of electrons that passes through the apertures of the cathode shield and the anode and down the microscope axis.

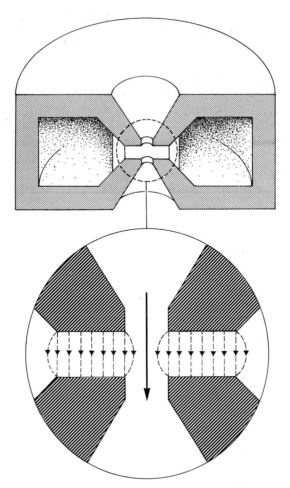

Figure 8. *A cut-away view of an electromagnetic lens, showing the cavity containing the wire coil surrounded by the soft iron casing. The detail (below) shows the position of the magnetic field, and the arrow indicates the direction of the electron path.*

The electromagnetic lenses and their arrangement

To focus the electron beam, electromagnetic lenses employ a strong axial magnetic field in the direction of the electron beam (see Figure 8). Each lens surrounds the microscope axis and consists of a wire coil held in a soft iron casing. The magnetic field is generated by passing a current through the coil. Since the focal length of the lens is determined by the strength of the magnetic field in the lens, focusing can be controlled by simply altering the current flowing through the coil.

In order to take full advantage of the resolving power of the objective lens there are three, or sometimes four, stages of magnification in an

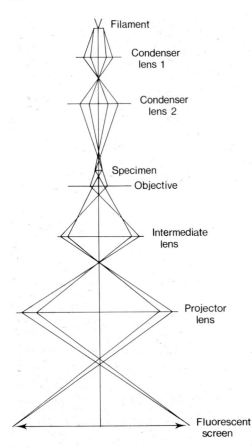

Filament

Condenser
lens 1

Condenser
lens 2

Specimen
Objective

Intermediate
lens

Projector
lens

Fluorescent
screen

Figure 9. *The path of electrons in an electron micro-scope. After A. W. Agar.*

electron microscope imaging system (Figure 9). Thus, as a rule, in addition to the objective and projector lenses, there are also 'intermediate' lenses which increase the range and flexibility of the magnifications that can be obtained.

In most instruments there are also two condenser lenses, because this arrangement produces a narrower and (for the specimen) less damaging illuminating beam.

The final image

Electrons are not 'seen' directly by the retina of the eye, and so, in a position equivalent to that of the light microscope ocular lens, there is a projector lens that produces a final image on a fluorescent viewing screen (Figures 10 and 11). The electrons impinge upon the phosphor crystals of the screen coating and induce them to fluoresce. Alternatively, since electrons also alter silver bromide crystals in a manner similar to photons of light, the projected image can be recorded photographically.

One important consideration in the design of the electron microscope is the fact that electrons are deviated when they collide with gas molecules, and they thus have a negligible penetrating power in air. The path

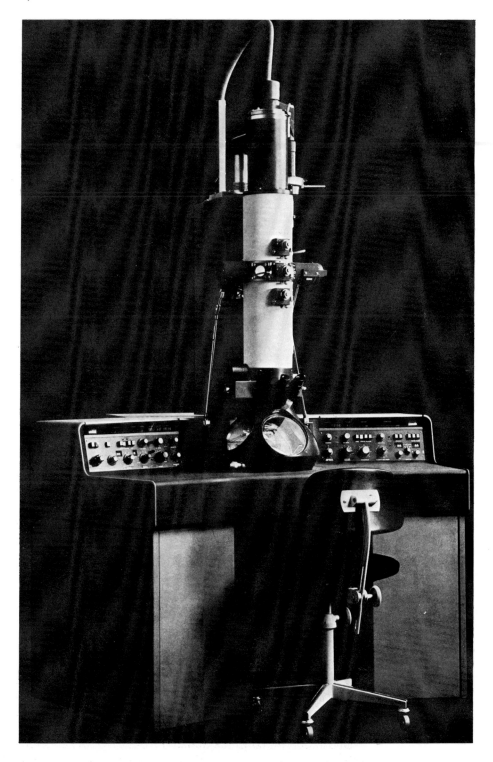

Figure 10. *The Philips EM 400 advanced research microscope. Courtesy of Philips Ltd., Eindhoven.*

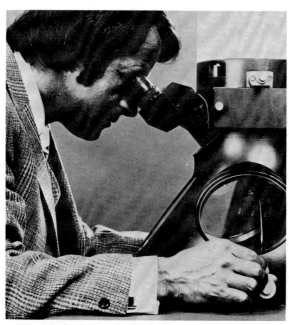

Figure 11. *Details of the Philips electron microscope in use – using the binocular magnifier to look through the viewing window and examine the image on the fluorescent screen. Courtesy of Philips Ltd., Eindhoven.*

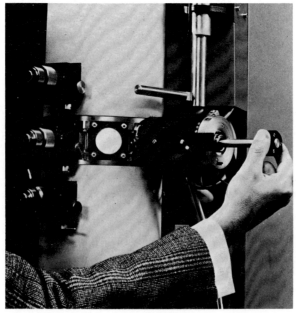

Figure 12. *Details of the Philips electron microscope in use – inserting the specimen carrier rod via the air lock. Courtesy of Philips Ltd., Eindhoven.*

of the electron beam must therefore be maintained under high vacuum (about 10^{-4} mmHg) and an elaborate system of vacuum pumps is necessary. For loading the specimen and photographic plates, air locks are provided (Figure 12).

SCANNING ELECTRON MICROSCOPES

These instruments are of a fundamentally different design to the transmission microscopes described above, since they were devised to examine the surfaces of solid objects. They use a fine beam of electrons to scan back and forth across the specimen surface and they then collect and process the low-energy, secondary electrons that are generated. To be examined in a scanning electron microscope, cell and tissue surfaces must, therefore, be preserved (i.e. fixed – see page 25), dried and coated with a thin, conducting, metallic film. (Gold and palladium heated to evaporation in a vacuum are most often used.) The secondary electrons arising from the specimen are collected by a scintillator crystal that converts each electron impact into a flash of light. Each of these light flashes in the crystal is then amplified by a photomultiplier tube and used to build the final image on a fluorescent screen. The scan of the beam used to form this image is driven in synchrony with that of the electron-exciting beam that scans the specimen, so the resulting image (formed in much the same way as the picture on a television cathode ray tube) is a faithful representation of the specimen surface as imaged by the output of secondary electrons. The magnification can be changed simply by altering the size of the area scanned by the scanning beam, thus:

$$\text{Magnification} = \frac{\text{Area of image (fixed)}}{\text{Area of object scanned (variable)}}$$

The scanning electron microscope is widely used for studying tissue surfaces because, as shown in Figure 13, it has considerable depth of focus. Unfortunately, however, because the size of the scanning beam is, for technical reasons, limited, the best resolution limit obtainable is only about 3 nm. Indeed, in most of the instruments presently available, it is only about 6 nm.

PHASE AND INTERFERENCE CONTRAST MICROSCOPES

Of the changes that occur in the properties of visible light, the eye detects only changes in wave amplitude (as changes in intensity) and changes

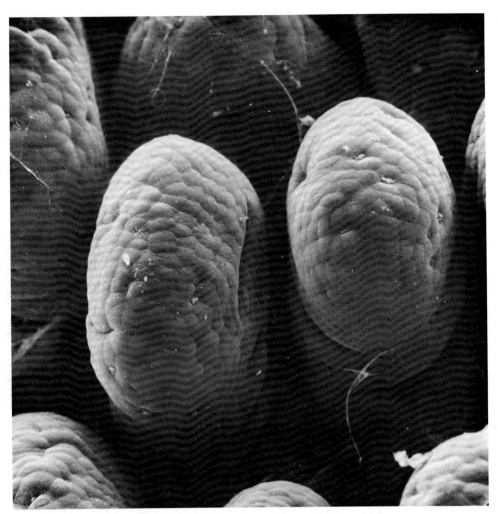

Figure 13. *A scanning electron micrograph of villi in the small intestine, showing the depth of focus that can be obtained. The polygonal outlines of the individual cells covering the villi are clearly seen. Scattered whisps of dried luminal content are also present. Magnification ×750.*
Courtesy of T. D. Allen, Paterson Laboratories, Christie Hospital and Holt Radium Institute, Manchester.

in wavelength (as changes in colour). Thin, transparent specimens like living cells or tissue sections do not significantly affect either of these wave properties, and thus, unless artificial contrast or colour can be introduced into them, they remain poorly resolved. For routine purposes (see below) artificial contrast can be introduced by staining preserved (dead) specimens, but for living preparations special optical systems are required. These systems exploit the fact that although thin, transparent specimens have little effect on the amplitude and wavelength of the light waves that pass through them, variations in refractive index and thickness within the specimen do affect the relative phase of these waves. Since waves of different phase can be induced to *interfere* and

1. Light waves may vary in (a) amplitude   or (b) wavelength

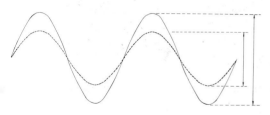
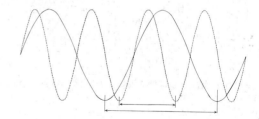

or (c) phase

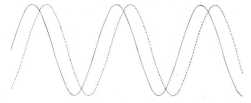

These waves are partially out of phase (by $\frac{1}{6}$ a wave-length)

2. When light waves arrive at the same point they interfere

3. Depending on the difference in phase, interference may result in an increase in the amplitude of the resultant wave (i.e. constructive interference)

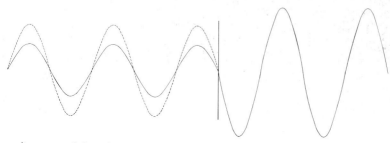

or it may result in a decrease in amplitude (i.e. destructive interference)

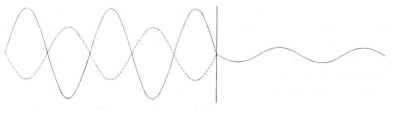

These waves are out of phase by $\frac{1}{2}$ a wavelength

Figure 14. *Some properties of light.*

cause changes in wave amplitude, these variations can be used to provide a visible 'amplitude image' of the specimen.

One of the most widely used of these specialized optical systems is phase contrast. This system will now be described in detail; to follow the account it may be helpful first to refer to Figure 14.

The design of the phase contrast microscope

1. Direct and diffracted rays

In a conventional microscope (as already indicated in Figure 4, but for this account refer to Figure 15), light rays emerging from a source via a field diaphragm (D) placed in the focal plane of a condenser lens (C) are transmitted to the objective lens (O) which focuses them in its rear focal plane (P_1) to form a real image in the image plane (I). These rays represent the path through the system of the *direct or unaltered* rays.

(In practice these direct rays can be demonstrated by setting up a conventional microscope for routine use—see appendix on page 22— and then, after removing the ocular [eyepiece] lens, closing the iris diaphragm of the substage condenser to a small, pinhole aperture. Without

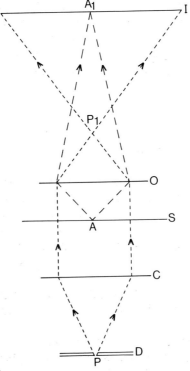

Figure 15. *The paths of direct and diffracted rays in the conventional light microscope.* – – – –, *direct rays;* ———, *diffracted rays. For details, see text.*

a specimen slide on the stage and looking down the barrel of the microscope, the focused rays of the condenser aperture are seen as a bright
spot in the back focal plane of the objective.)

Many of the light rays passing through a specimen will, however,
be diffracted and (as indicated by rays arising at A and focusing at A_1)
they will be focused in the image plane (I). At the level of the rear focal
plane of the objective lens (P_1), these diffracted rays, unlike the direct
rays, are therefore diffuse and unfocused.

(In practice this can be demonstrated by setting up a microscope,
as before, removing the ocular lens and, having observed the bright spot
of focused direct rays, putting a specimen slide on the stage. The bright
spot of the direct rays becomes surrounded by a halo of diffuse light;
these are the rays diffracted by the specimen.)

2. Changes in phase of direct and diffracted rays

When light rays are transmitted through a thin, transparent specimen
such as a cell, variations in optical density (refractive index × thickness)
within the specimen alter their phase. These variations most affect
the phase of the diffracted rays, and in practice there is, on average,
a difference in phase of about one-quarter of a wavelength between
the diffracted and direct rays emerging from the specimen. Although
these changes occur in all conventional microscope systems they are
not normally visible, because this degree of phase alteration is insufficient to affect the amplitude of the waves forming the final image
(the intensity of the direct rays of this image is, in any case, normally
overwhelming).

3. Separation of direct and diffracted rays

The arrangement of the optical components in the phase contrast microscope is designed to obtain a partial separation of the direct and diffracted rays in the rear focal plane of the objective lens. When this
separation is coupled with a further contrived increase in the difference
in phase between these rays, the constructive and destructive interference that results is sufficient to cause a visible change in wave amplitude.

In most phase contrast microscopes the partial separation of the direct
and diffracted rays is accomplished by inserting a clear annulus in the
focal plane of the condenser (D in Figures 15 and 16). The direct rays
then focus with the distribution of their source, that is, the annulus, in
the rear focal plane of the objective lens. (It should be remembered that
in this focal plane the diffracted rays are out of focus, and they are thus
not confined to the image of the annulus.)

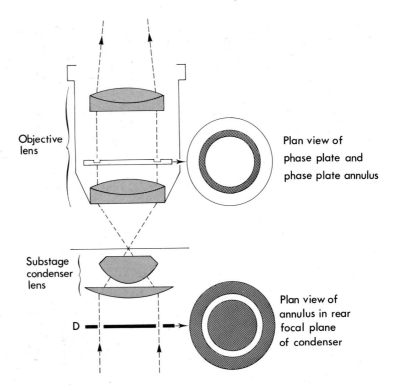

Objective lens

Plan view of phase plate and phase plate annulus

Substage condenser lens

D

Plan view of annulus in rear focal plane of condenser

Figure 16. *The arrangement of the substage condenser annulus and the phase plate in a phase contrast microscope. The condenser annulus (shown in plan view on the right) produces a hollow cone of rays that focuses on the specimen and provides an image of the annulus in the rear focal plane of the objective lens. In this plane the phase plate annulus (shown in plan view on the right) is positioned so that it coincides with the image of the condenser annulus.*

With the phase plate annulus shown in this diagram, the 'direct' rays (————) passing through the annulus will be retarded relatively less than those that pass through the thicker surrounding plate. A 'positive' phase contrast image will therefore be produced.

4. The phase plate annulus

For 'positive' phase contrast* (see Figure 16) the phase of the wavelength of the direct rays is 'advanced' an additional one-quarter of a wavelength in the rear focal plane of the objective lens by incorporating a phase plate annulus in this plane (P_1 in Figure 15), so that it precisely coincides with the annular distribution of the direct rays. The glass of the annulus is thinner than that of the surrounding phase plate so that the rays that pass through it (mostly direct rays) are less retarded (by about one-quarter of a wavelength) than those rays that pass through

---

* For 'negative' phase contrast the phase annulus is thicker than the surrounding phase plate and the direct rays are thus retarded relative to the diffracted rays.

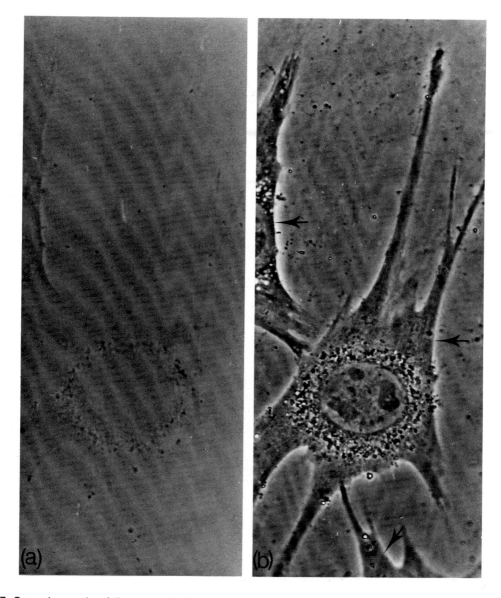

Figure 17. *Two micrographs of the same cell taken (a) with conventional optics and (b) with phase contrast. The arrows indicate the halo effect that results from an incomplete separation of the direct and diffracted rays. Magnification × 4000.*

the surrounding plate (mostly diffracted rays). The total average shift in phase between the direct and diffracted rays is therefore now about one-half of a wavelength.

In the rear focal plane of the objective the intensity of the direct rays (which would otherwise be excessive) is also reduced by coating the phase plate annulus with a thin metal film.

5. The final image

In the image plane (I in Figure 15), the direct rays, reduced in intensity and now, on average, one-half a wavelength out of phase with the dif-fracted rays, interfere with the diffracted rays. Constructive and destruc-tive interference results, giving rise to an *amplitude* image. In positive phase contrast those components in the specimen with a refractive index relatively greater than their surroundings appear in shades of grey (de-structive interference), while those with a refractive index relatively lower than their surroundings have a high, bright intensity (constructive interference).

Unfortunately, at the edge of refractile components, secondary inter-ference arises (see below) and the boundaries of these components are therefore characteristically ringed by a bright halo that obscures fine detail. Nevertheless, as shown in Figure 17, phase contrast microscopy still allows amplitude images of considerable detail to be obtained.

The design of interference microscopes

The secondary interference that occurs in the phase contrast microscope arises primarily because the simple arrangement of a condenser annulus image and a coincident objective phase plate is unable to separate direct and diffracted rays completely. In interference microscopes this difficulty is overcome because the direct and diffracted rays are completely separated using a beam-splitting device to conduct the direct light rays (along an equivalent path length) around rather than through the speci-men. In some instruments it is possible to vary the relative phase and amplitude of the direct rays so that the degree of alteration in the dif-fracted rays caused by a particular cell component can be measured. The refractive index of the component can then be estimated and its relative mass (i.e. its concentration) quantified.

APPENDIX TO THE MICROSCOPY SECTION

Setting up a microscope with an external light source

1. Refer to Figure 18 and note the labelled components.
2. Tilt the body of the microscope towards the horizontal and, with the mirror facing you, examine the substage condenser. Note the follow-ing:
   (a) The iris diaphragm is opened and closed by a small lever — leave the diaphragm open.
   (b) The condenser can be moved up and down on a rack and pinion slide — leave it racked full up, almost level with the specimen stage.

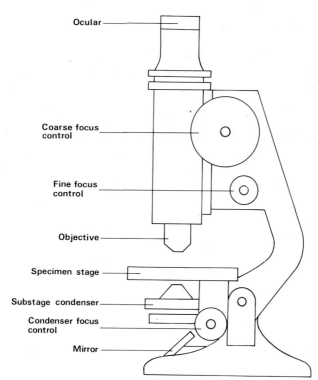

Ocular

Coarse focus control

Fine focus control

Objective

Specimen stage

Substage condenser

Condenser focus control

Mirror

Figure 18. *The light microscope.*

3. Return the body of the microscope to the upright position.
4. Place a stained microscope specimen slide on the stage, select the low-power (×10) objective, roughly align the light source and the mirror and, looking through the ocular, use the coarse focus knob to focus the specimen. The light source should be about 20 cm from the mirror and only the plane surface of the mirror should be used (the concave surface is used only with specialized 'ultra low-power' objectives).
5. Close the iris diaphragm of the substage condenser; looking through the ocular you should see a central bright spot and, as you open the iris, the spot should open up and spread to illuminate the field evenly. (If the spot is off-centre the condenser needs to be realigned within its housing.)
6. Open the condenser iris diaphragm and realign the light source and mirror so that the field of view is evenly illuminated.
7. To focus the condenser put the tip of a pencil against the centre of the lamp and, looking through the ocular, gently rack the condenser up and down until the image of the pencil tip is seen in sharp focus. (You will find that this is so when the condenser is almost fully racked up; if the image of the light bulb is disconcerting, the condenser can be fractionally defocused.)

8. Check that the specimen is in focus and then remove the ocular. From some distance above, look down into the barrel of the microscope at the back of the objective lens. Slowly open and close the condenser iris diaphragm and then set it so that about two-thirds of the lens is illuminated; this is the correct setting for this objective lens. Replace the ocular and examine the field. The microscope is now correctly set up for use with the low-power lens.

9. Switch to the ×40 objective and, using the fine focus control, refocus. Remove the ocular and examine the iris aperture as before. You will find that the setting no longer illuminates two-thirds of the field (because this objective has a higher NA) and the iris will thus need opening to obtain the correct (two-thirds) setting. This readjustment needs to be made every time the objective lens is changed.

Note: With this system the light intensity should only be adjusted by moving the lamp housing towards or away from the microscope.

Microscopes with built-in illumination

1. The light intensity is controlled by a switch.
2. Proceed as above except that in focusing the condenser the pencil should be placed on top of the lens, covering the light source.

Other modifications

1. The substage condenser may have a flip-top lens designed for use with high-power objectives. This lens should be flipped out of position when low-power objectives are aligned.
2. There may be a field diaphragm fitted to the illumination source. Partially closed, the edges of this diaphragm can be used (instead of a pencil) for focusing the substage condenser. At its correct, final setting this diaphragm should just clear the field of view.

2

PREPARATION OF THE SPECIMEN FOR MICROSCOPY — ROUTINE METHODS

Because conventional light and electron microscopes are transmission microscopes, they depend for their illumination upon light rays or electrons which pass *through* the specimen. Thin, wafer-like preparations are therefore required, and for routine, non-specialist purposes these are obtained by cutting preserved and embedded blocks of tissue into thin sections.

THE PRESERVATION OF CELLULAR COMPONENTS — FIXATION

Once they are removed from their normal physiological environment, the morphological interrelationships within and between cells can be faithfully preserved only if their main structural components are efficiently stabilized and if autolysis is prevented. This process is called 'fixation', and it can be achieved either by physical or chemical methods. The most common method of physical fixation relies upon rapid freezing; chemical fixation, on the other hand, requires reagents that precipitate, denature and/or cross-link the structural components of the cell.

Historically, the development and use of chemical fixatives has been essentially an empirical process, and for routine histological purposes a variety of reagents is now available. However, in recent years, when the increased resolution obtained by the electron microscope has required optimum preservation of cellular and subcellular components (and applied techniques such as enzyme histochemistry have placed additional demands on the fixation process), only a few of these fixatives have been able to meet the more stringent requirements. These reagents have been satisfactory mainly because they preserve the membranous substructure of the cell. They do so primarily by stabilizing the protein components of the membranes; alternatively (or in addition) they may cross-link the membrane lipids.

Proteins. Linear polymers of amino acids that may be twisted, pleated or folded.

Tissue proteins frequently contain reactive sites, such as the hydroxyl, carboxyl and carbonyl groups shown in Formula 1, and it is these that

Formula 1. *Hydroxyl, carboxyl and carbonyl groups present on two amino acids – glutamic acid and tyrosine – in a polypeptide chain.*

can interact and be stabilized with fixative reagents. However, in addition and of special note in the stabilization of tissue proteins is the cross-linking of protein polypeptide chains that can be achieved with aldehydes like formaldehyde and glutaraldehyde. This cross-linking involves the free amino groups of the proteins and can occur as shown in Formula 2. Formaldehyde is an important fixative in routine histology (and often

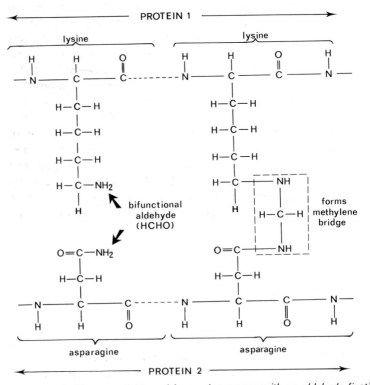

Formula 2. *The cross-linking of free amino groups with an aldehyde fixative.*

essential in histochemistry), while glutaraldehyde, alone or in combination with formaldehyde, is widely used in electron microscopy.

When using osmium tetroxide as a fixative, stability is achieved largely by cross-linking the tissue lipids. This fixative, which also has some reactivity with certain of the amino acids in tissue proteins, probably acts primarily by cross-linking the chains of unsaturated fatty acids, as shown in Formula 3.

Lipids. A heterogeneous class of water-insoluble compounds extractable by non-polar solvents like chloroform and ether.

Formula 3. *The cross-linking of unsaturated fatty acid chains with osmium tetroxide.*

To ensure the initial entry of a fixative reagent into the tissue causes as little premature disturbance as possible, it is normally buffered to a physiological pH (7.4) and maintained at a suitable osmolality (about 300 milliosmol). For most light microscope studies the aldehydes are used alone, but for electron microscopy they are normally employed in conjunction with osmium tetroxide, since this heavy metal will fix additional tissue components and, by becoming incorporated into cellular membranes, it also enhances their electron-opacity (see below). A sufficiently fast and efficient penetration by these fixatives is often a problem because, by cross-linking the tissue components, they may hinder their own progress into the deeper regions of the tissue. This difficulty may sometimes be alleviated by perfusing the vasculature of an organ with aldehyde immediately after death, but the most usual fixation method involves quick removal from the body, immersion in the fixative, and simple dicing of the tissue into small blocks (about 1 mm³). Clearly, difficulties of orientation and sampling errors must be kept in mind when these small random tissue fragments are later examined in the microscope.

EMBEDDING AND TISSUE SECTIONING

For tissue blocks to be cut into thin sections they need to be rigidly supported. Frozen tissue blocks are adequately supported by their contained ice and require no further treatment, but chemically fixed tissues must be surrounded and penetrated by a hard and mechanically strong supporting ('embedding') medium. The most widely used embedding media are either waxes or epoxy resins, and since most of these are immiscible with water, embedding procedures usually begin with the removal of water from the fixed tissue. The dehydration procedure simply entails the transfer of the fixed tissue blocks through a graded series of alcohol/water solutions (e.g. 50, 75, 90, 95 and 100 per cent ethanol), a period for equilibration between the alcohol and the tissue content being allowed in each solution (usually of some minutes). On reaching and becoming equilibrated with 100 per cent alcohol, the fully dehydrated tissue is ready to be embedded. It cannot be directly infiltrated with embedding media, however, since the most widely used media are either immiscible with alcohol or otherwise only poorly penetrate alcohol-impregnated tissue. It is therefore necessary to transfer the dehydrated tissue to an intermediary reagent, such as xylene or propylene oxide, that is miscible with both alcohol and the embedding medium. The effect of these intermediary reagents upon tissues is to make them transparent rather than translucent, and they are therefore often called 'clearing' reagents.

Finally, the tissue is infiltrated with embedding medium. Media used for embedding have a liquid and a solid phase, which allows for their liquid infiltration throughout the tissue interstices before a period of setting and hardening. Satisfactory embedding media penetrate even the densest components of the cell and on hardening they neither shrink nor distort. For light microscopy, wax, infiltrated when molten (at about 50°C) and hardened by cooling, usually completes the embedding process. For electron microscopy the requirements for sectioning are more stringent and the embedding medium must also serve to support the tissue section within the microscope. It is therefore necessary to use the mechanically stronger and more heat-stable epoxy resins such as Epon and Araldite. These resins are infiltrated into the tissue in an unpolymerized, liquid phase over a period of hours, before being polymerized and hardened by heating to about 60°C for some days.

Sections are cut from embedded or frozen tissues on an instrument known as a microtome. In its most basic form, a microtome consists of a knife, a specimen holder and an advance mechanism. These components are arranged so that, following the incremental advance of the tissue block, it traverses the fixed knife-edge and a section is cut from the advancing face (Figure 19). After each section is cut it remains attached to the knife-edge along its rear margin. The specimen is then again advanced and the cutting process is repeated. As each section emerges from the knife-edge it adheres along the length of its advancing edge to the rear edge of its predecessor and in this way a ribbon of sections arises.

Sections cut from frozen tissue blocks are not normally thinner than 7 to 8 μm thick and must be cut at low temperature ($-20°$C) on a refrigerated microtome called a 'cryostat'. Since embedding and often fixation can be dispensed with, cryostat sections have the advantage of being quick and easy. However, on thawing, frozen tissues suffer ice-

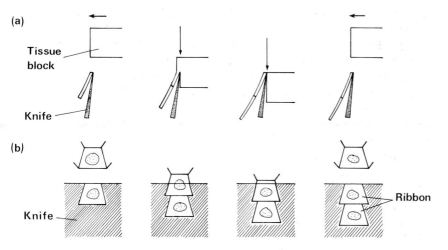

Figure 19. *A ribbon of sections being cut on a microtome:* (a) *seen from the side,* (b) *seen from the front.*

crystal damage and their preservation, though adequate for many diagnostic purposes, is generally inferior.

Wax sections are routinely cut 3 to 5 μm thick and readily form ribbons, which for further processing are usually cut into 5- to 10-section segments; they are then floated over a pool of adhesive albumin onto a glass microscope slide. With gentle heating the aqueous albumin gradually evaporates and the sections flatten and then stick.

For conventional transmission electron microscopy, epoxy resin sections are usually cut as thin as 80 nm. This is because of the poor penetrating power of the electron beam. These so-called 'ultra-thin' sections are cut on ultramicrotomes – microtomes engineered to fine tolerances and which (because steel knives cannot be made sufficiently sharp) use carefully fractured glass or diamond knives as the cutting edge (Figure 20). Cutting ultra-thin sections is a demanding process and usually the area of block face being sectioned must be reduced to about 0.1 mm². As the fragile ribbon of sections is produced it floats out and is supported over a trough of water held so that the meniscus of the water just touches the knife-edge (Figure 21).

The considerations that dictate the need for ultra-thin sections also mean, of course, that a matrix other than glass must be used to support the sections in the microscope. The ribbons of ultra-thin section are therefore lifted from the water surface and mounted instead on fine copper grids. As shown in Figure 21, the sections on the grids are exposed and available for examination where they are draped across the holes, while they receive their necessary support from the intervening metal cross-pieces.

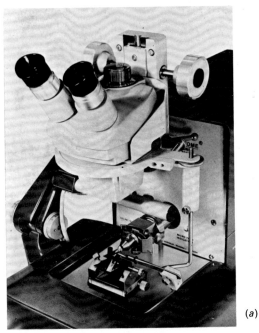

(a)

Figure 20. (a) *A view of the cutting area of an ultra-microtome (LKB Ultratone III). (b) A view through the binocular microscope showing the embedded tissue block approaching the knife-edge. The specimen support grid is held in readiness in the tip of the forceps.*
Courtesy of LKB Instruments Ltd., Selsdon, Surrey.

(b)

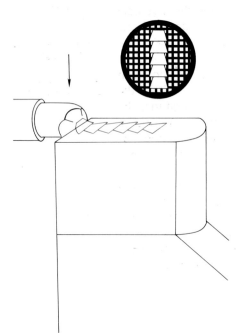

Figure 21. *A ribbon of sections being cut on an ultra-microtome. The sections emerging at the knife-edge float out on the surface of the water in the knife trough. From there they are picked up on the copper specimen grid shown above.*

PREPARING SECTIONS FOR LIGHT MICROSCOPY

Earlier, in the section dealing with phase and interference contrast microscopes, it was emphasized that, unless differential contrast or colour can be introduced into the preparation, the components of cells and tissues are difficult to see with the conventional light microscope. For routine light microscopy, therefore, tissue sections are usually stained with reagents which bind preferentially to one or more components. Many of the available histological stains were originally produced as dyes in the textile industry and their use in histology has been developed largely by a process of trial and error. As a rule they are used in aqueous solution, which means that, while cryostat sections can be stained directly, wax sections must first have the hydrophobic embedding medium removed and water re-introduced into the tissue. Wax removal and rehydration is achieved simply by immersing the slide bearing the sections in xylene and then transferring it sequentially down a graded series of alcohols to water. The section is then ready for staining.

Histological stains and staining

The binding of many histological stains by tissue components depends on electrostatic (salt) linkages being formed between the stain and the ionizable radicals of the tissue components. These stains are either anionic, and bind to cationic components, or vice versa. The overall charge of the stained tissue component will depend upon the algebraic sum of its positive (cationic) and negative (anionic) charges and this may be influenced by the pH of the stain solution. Thus, in proteins the constituent amino acids tend to be cationic at low pH (with their free amino groups ionized), while at a higher pH they become anionic (with their carboxyl groups ionized). Additional ionizable groups within the molecule (such as the extra amino groups of lysine and arginine, the phenolic hydroxyl group of tyrosine, and the phosphate groups of nucleic acids) will also contribute, so that at the given pH, and depending upon its ionizable constituent groups, a tissue component may be either anionic or cationic.

At pH 5 to 6 (the usual range employed for staining), the predominant components of the ground cytoplasm of cells are cationic, as is also the matrix of the red blood cell (due to the lysine content of haemoglobin). This means that these components bind preferentially to anionic stains such as eosin (Formula 4) and orange G. Chromatin and the ribonucleic acid components of cytoplasmic ribosomes, on the other hand, are anionic in this pH range, binding cationic stains such as methylene blue and toluidine blue.

Electrostatic link. When a stationary electrically charged particle attracts another particle of opposite charge.

'Anionic stains' see *Basic stains*[*]

'Cationic stains' see *Acidic stains*

Cytoplasmic basophilia

---

[*] For definitions of the italicized marginal entries, see the Glossary.

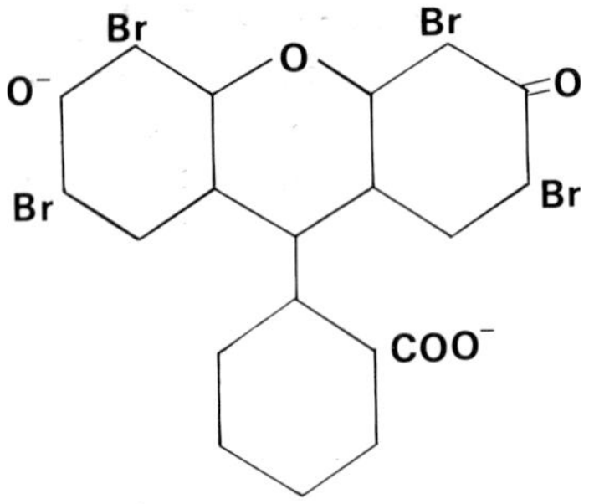

Formula 4. *Eosin Y.*

Haematoxylin and eosin

Not all staining reactions depend upon a direct anionic–cationic reaction, however, and an important exception to the rule is that of haematein (Formula 5), the active component of haematoxylin.

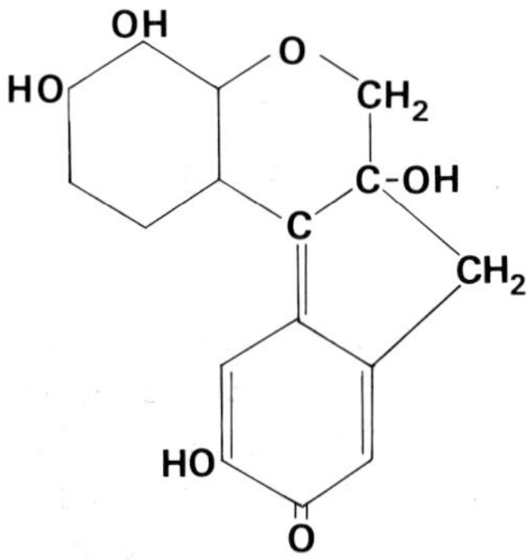

Formula 5. *Haematein.*

Mordant and
Lake are terms
used in the dyeing
industry.

A staining solution of haematoxylin includes haematein (a negatively charged oxidized form of haematoxylin), excess potash alum $(K_2SO_4Al_2(SO_4)_3.24H_2O$ – usually called the 'mordant') and acetic acid. For staining, a tissue section is treated with this solution before being rinsed in alkalinized water (tap-water in suitable districts). The rinsing procedure neutralizes the effect of the acetic acid and raises the pH. As a result the haematein complexes with the alum mordant to form a blue-coloured complex known as a 'lake' (the original haematoxylin staining solution has a reddish, dun colour). The lake, unlike haematein, is positively charged (cationic) and thus binds to negatively charged tissue components. In staining with haematoxylin, therefore, the mordant complex, during the alkaline rinse (usually known as 'blueing'), induces a negatively charged staining reagent to bind to negatively charged tissue components. To produce the widely used 'H and E' preparation, haematoxylin staining is followed by eosin, a negatively charged dye that binds preferentially to the cationic, positively charged components of the tissue. The result: nuclear chromatin, blue; ground cytoplasm, red.

Post-staining treatment

Once the staining of the tissue section is complete, it is usual for routine purposes to make the preparation permanent by mounting a thin glass coverslip over the surface. For this purpose mounting media are used that, on drying, harden and provide a clear, optically acceptable medium between the coverslip and the section. Most mounting media are either natural or synthetic resins. They are immiscible with water and mix only poorly with absolute ethanol, so that, as in embedding, it is necessary to dehydrate the preparation using a graded series of alcohols and employ xylene (which is miscible with both absolute alcohol and most mounting media) as an intermediary. Following these preparative procedures, the fixed, stained section is finally ready for examination under the microscope.

PREPARING SECTIONS FOR ELECTRON MICROSCOPY

The formation of an image in the electron microscope depends upon the beam of electrons emerging from the electron gun, passing down the axis of the microscope through the ultra-thin section, and impinging upon either a fluorescent screen or the emulsion of a photographic plate. The atomic and molecular constituents within the section will scatter and deflect those electrons which come in contact with them, often to the extent that the electrons are lost from the beam. The supporting epoxy resin does not, however, significantly affect the transmitted electrons, and thus electrons passing through resin alone will remain within the microscope axis and reach the fluorescent screen. The extent to which electrons will be scattered by any cellular component within the section depends upon the number of atoms per unit volume and the size of these atoms, and it follows that the more large atoms there are within a sectioned component the larger will be the number of electrons removed from the beam and the darker the image. The size of an atom and thus its ability to scatter electrons is indicated by its atomic number. Unfortunately the common atomic constituents of cells, such as carbon, hydrogen and oxygen, are all of low atomic number, and so, to provide adequate contrast in the electron microscope, it is necessary to 'stain' cellular components differentially with heavy metal salts of high atomic number.

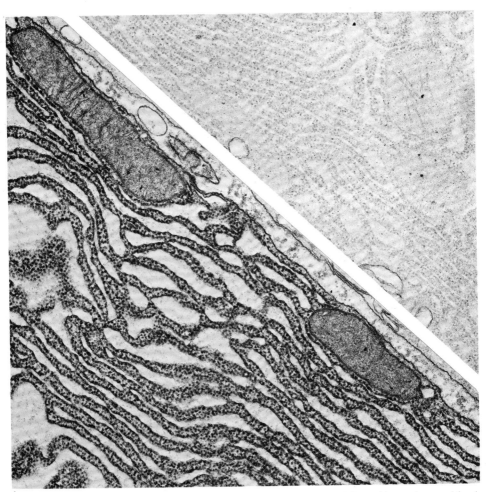

Figure 22. *The same tissue section before (upper right) and after staining with uranium and lead salts.*

'Staining' is achieved by simply immersing the sections, mounted on their supporting grid, in a solution of the appropriate metal salt. It has been found that the salts of uranium readily penetrate the epoxy resin of the section and bind preferentially to nucleic acids and proteins, while lead salts penetrate and are taken up strongly by lipid components. Used together, these electron 'stains' provide a most satisfactory image of tissue sections, since they increase the contrast of cellular membranes (and to a varying extent other cellular components of interest) while they leave the epoxy resin matrix unstained (Figure 22).

3

PREPARATION OF THE SPECIMEN FOR MICROSCOPY – APPLIED TECHNIQUES

HISTOCHEMISTRY

Histochemical techniques include a diverse group of methods primarily concerned with identifying specific chemical components or reactive groups within a tissue section. For any histochemical technique to be acceptable it must be based upon a predictable and fairly specific chemical reaction and be able to produce an identifiable (by contrast, colour or electron-opacity) reaction product within the tissue. In addition, the morphology of the tissue should remain adequately preserved and the reaction product formed must remain at the intracellular location in which it is initially formed.

A routine technique, the periodic acid–Schiff (PAS) method, will serve to illustrate the kind of approach adopted in histochemistry. This technique, which is widely used in diagnostic pathology, is used to identify certain tissue carbohydrates. Before discussing the method it will help to categorize briefly the tissue carbohydrates that need to be considered.

The carbohydrates of animal tissues, with the exception of glycogen (a polysaccharide containing only sugar residues), all consist of hexose-related sugars attached to either proteins or lipids. The carbohydrate/protein molecules include the mucopolysaccharides (glucosaminoglycans), in which the carbohydrates are well represented (less than 40 per cent), and glycoproteins, which contain only about 4 per cent or less carbohydrate. Neutral mucopolysaccharides are the major constituents of mucous secretions, while acid mucopolysaccharides are widely represented by the hyaluronic acid of connective tissues. Glycoproteins contain hexoses (such as glucose, galactose and mannose) and hexosamines (such as glucosamine), and they are functionally very important. As major components of the plasma membrane of the cell, they are, of course, ubiquitous, while as tissue antigens, immunoglobulins, serum proteins and pituitary hormones they are often concerned with a discriminating functional specificity.

Antigen

'Immunoglobulin'
see *Antibodies*

Hormones

The PAS procedure, which is routinely carried out on wax sections, begins with an initial periodic acid oxidation step. This mild oxidant breaks the C–C linkage in $1:2$ glycol groups, converting them to aldehydes (Formula 6), and has the special advantage that its oxidation

Formula 6. *The production of aldehydes from glycoproteins by PAS oxidation.*

proceeds as far as the production of the aldehyde and no further. It does not affect the aldehydes themselves. Since lipid/carbohydrate molecules are normally removed from the tissue during wax embedding, and the hexosamine and hexuronic acid groups of acid mucopolysaccharides are apparently unaffected by periodic acid, only the hexoses and hexosamines of glycogen, glycoproteins and neutral mucopolysaccharides are oxidized to aldehydes.

The aldehydes can be identified by treating the section with Schiff's reagent, a basic fuchsin dye decolorized with sulphurous acid, which undergoes a specific but complex reaction with aldehydes to produce a brilliant magenta-coloured dye.

As with any histochemical method, the specificity of a positive reaction is always tested using control preparations. Thus, by omitting the periodic acid oxidation step from the PAS method, tissue aldehydes other than those derived from hexoses and hexosamines can be identified. Also, since glycogen can be selectively removed from tissue sections by diastase (beta amylase, an enzyme that depolymerizes this complex carbohydrate into readily soluble and thus readily removed monosaccharides) the presence of glycogen can be tested for by comparing diastase-pretreated with diastase-untreated PAS-stained sections (Figure 23).

Enzyme histochemistry in light microscopy

Most of the histochemical methods by which tissue enzymes can be identified rely upon the activity of the enzyme itself for their specificity.

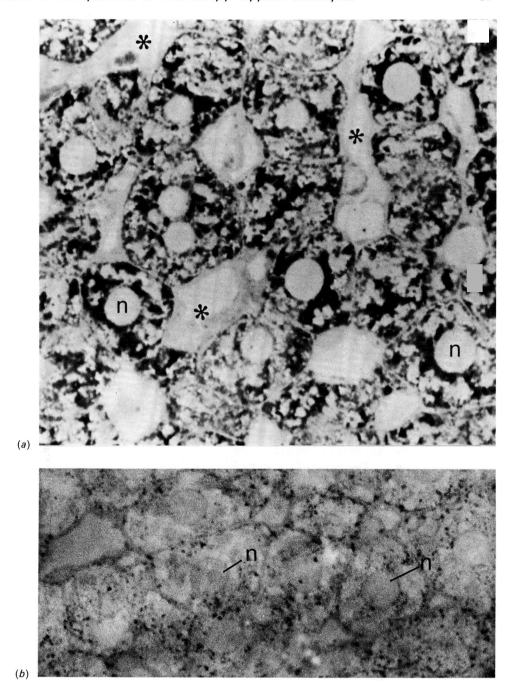

(a)

(b)

Figure 23. (a) A liver section stained with the PAS technique. The technique stains dense, coarse aggregates within the cytoplasm. The lumina of the blood vessels (*) and the nuclei (n) are indicated. (b) A similar preparation to (a) but pretreated with diastase. The coarse aggregated material is no longer seen, and it can be concluded that the positive reaction seen in (a) is due to glycogen. Both magnifications ×800.

This means that the activity of the enzyme within the tissue needs to be protected during section preparation. Fortunately, although it is occasionally necessary to avoid any denaturation by using unfixed cryostat sections, many enzymes will adequately survive aldehyde (and especially formaldehyde) fixation. In unfixed sections, diffusion of the enzyme and its reaction products away from their original location within the tissue can be a major disadvantage.

The lead salt method for acid phosphatase

One of the most widely used enzyme histochemical methods is the lead salt procedure for acid phosphatases. This is especially informative because it identifies an enzyme diagnostic of a specific cell organelle, the lysosome (see page 173). It can be applied at both the light and electron microscope levels.

For light microscopy a two-step procedure is used. In the first step, the fixed sections are incubated in a 'substrate' bath in which the tissue-bound enzyme catalyses the formation of an insoluble reaction product.

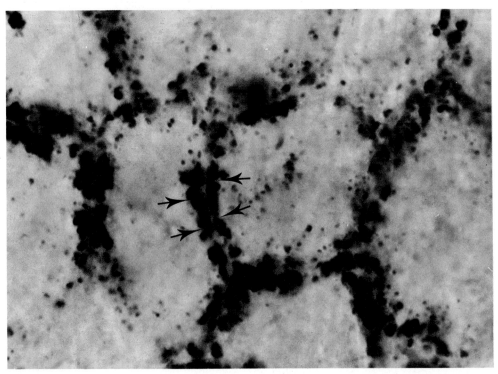

Figure 24. *Acid phosphatase localized using the lead salt method described in the text. The dense reaction product is confined to small particles arrayed predominantly along the cell boundaries (arrows). These particles are lysosomes. Magnification ×1600.*

In the second, a 'developing' bath is used to transform this primary reaction product into a coloured, secondary reaction product. The substrate bath contains a phosphoric acid monoester (sodium β-glycerophosphate) as the substrate, a buffer to maintain an acid pH (5.5), and a 'capture agent' in the form of lead ions. The acid phosphatases in the tissue hydrolyse the substrate to yield phosphate ions (plus sodium glycerate), and the lead ions immediately react with the phosphate to form insoluble lead phosphate (the primary reaction product). Treatment of these sections with ammonium sulphide converts the lead phosphate to lead sulphide, and this salt appears as a non-diffusible, black precipitate within the section at the site of the acid phosphatase (Figure 24).

$$PbPO_4 + (NH_4)_2 S \rightarrow PbS$$

The specificity of the procedure is usually checked by omitting the β-glycerophosphate from the substrate bath, or by adding a known inhibitor of the enzyme to the complete substrate mixture; for this particular enzyme, sodium fluoride is often used.

Enzyme histochemistry in electron microscopy

Where an electron-opaque reaction product is produced, a histochemical technique can often be applied at the electron microscope level. For example, since lead phosphate, the primary reaction product of the acid phosphatase technique, is intensely electron-opaque, the method has been adapted and used widely in fine-structural studies on the form and distribution of lysosomes. Satisfactory tissue preservation and adequate enzyme activity (about 20 per cent of the initial activity) are obtained with aldehyde fixatives, and thus 'frozen' or otherwise 'cold' sectioning methods for fixed tissue can be used to provide sections for incubation in the substrate bath. For electron microscope examination, however, subsequent processing must be concerned with providing adequately supported ultra-thin sections. The reacted tissue sections are therefore dehydrated, and embedded in epoxy resin as if they were tissue blocks. They can then be sectioned on an ultramicrotome and prepared finally for the electron microscope (see Figure 121).

Haem enzyme histochemistry

Some compounds act as hydrogen donors (i.e. they can be oxidized), and when they do so they produce insoluble and coloured dyes. For example, alpha-naphthol and dimethyl-*p*-phenylenediamine can be oxidized by the respiratory pigment cytochrome c to provide the coloured dye, indophenol blue, as shown in Formula 7.

This reaction has been used histochemically to identify the distribution of 'oxidase' systems (especially those of the cytochromes – see page

Oxidation. Addition of oxygen or removal of hydrogen. Also any reaction in which an atom loses electrons.

Formula 7. *The formation of indophenol blue by oxidizing alpha-naphthol and dimethyl-p-phenylenediamine with cytochrome c.*

200) within cells. A similar approach has also been developed for identifying peroxidases, enzymes that catalyse electron transfer in the presence of hydrogen peroxide:

$$AH_2 + H_2O_2 \xrightarrow[\text{peroxidase}]{} A + 2H_2O$$

Provided the oxidation product (A) is insoluble, and can be made coloured or otherwise identifiable, the distribution of peroxidase can then be demonstrated. Fortunately, the aromatic, diamine,3',3'-diaminobenzidine (DAB), provides this kind of oxidation product; in the presence of peroxidases and H_2O_2 it forms a cyclized polymer, which, when treated with osmium tetroxide (OsO_4), gives a complex product known as osmium black (Formula 8). Since osmium black is both coloured and electron-opaque, the technique can be applied at both light and electron microscope levels.

In fine-structural studies, the DAB reaction has been widely used for locating both endogenous and exogenous peroxidases. This is partly because peroxidases survive aldehyde fixation particularly well, and partly because only relatively few peroxidase molecules are required to produce a significant amount of reaction product. With this method it has been possible, for example, to identify endogenous peroxidases in the cisternae of the rough endoplasmic reticulum, a location in which they are probably present at very low concentrations (see Figure 100).

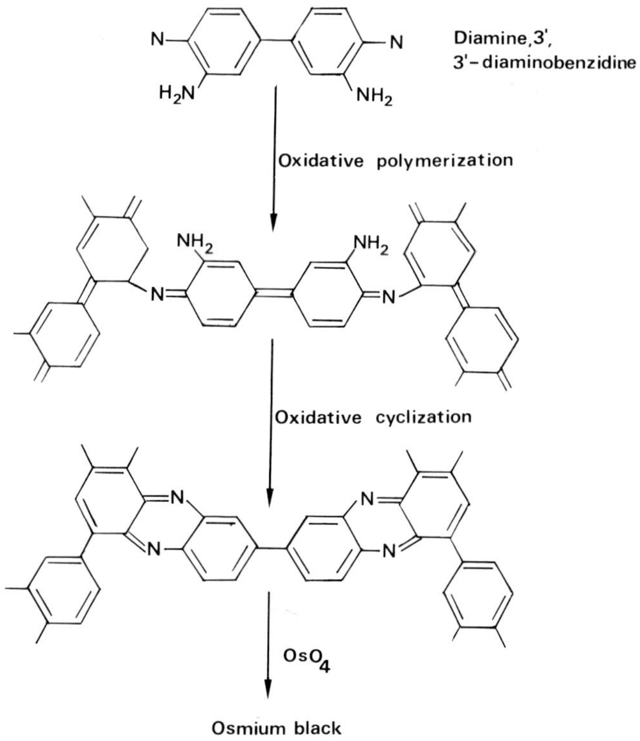

Diamine,3',
3'-diaminobenzidine

Oxidative polymerization

Oxidative cyclization

OsO_4

Osmium black

Formula 8. *The production of osmium black.*

These advantages of peroxidase histochemistry have been widely exploited because of the availability of a relatively pure and fairly inexpensive enzyme preparation – horseradish peroxidase. This enzyme (mol. wt. about 40 000) has been used extensively as a microscopically identifiable tracer in a wide variety of studies. It has been used, for example, to examine the permeability of intercellular junctions and tissue barriers like the capillary wall. It has also been used in many immunocytochemical studies to label antibody molecules (see below).

IMMUNOCYTOCHEMISTRY

In this approach, particular cell constituents are located within a tissue section by exploiting their capacity as 'antigens' to bind selectively to specific 'antibodies'. An antigen is a substance which, when injected into an animal, induces the animal's immune system to produce an antibody. Antibodies, which belong to a class of serum glycoproteins, perform an important role in the body's defence system, because they are able to recognize, bind, and precipitate (usually) their antigens with a very high degree of specificity. Thus, provided the cellular constituent to be located can be extracted from the tissue (or otherwise obtained

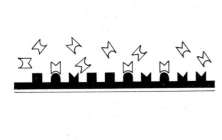

1. Antigen sites are available on the section surface.

2. Incubate the tissue with rabbit antibody specific for a particular antigen.

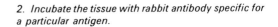

3. Wash away the excess antibody — only specifically bound antibody remains.

4. Incubate with goat anti-rabbit antibody labelled with fluorescein — it binds to the rabbit antibody because this is its antigen.

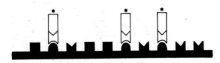

5. Wash away excess labelled antibody — only specifically bound antibody remains. Fluorescent label indicates the location of the tissue antigen.

Figure 25. *Fluorescence immunocytochemistry — the indirect method.*

pure) in sufficient quantity, it may be injected as an antigen into another animal species and so raise a specific antibody. This antibody can then be attached to an identifiable marker or label and will specifically bind (i.e. 'stain') its antigens when they are located within a tissue section.

The essential requirements for this technique are: (a) pure antigen, (b) a specific antibody that binds avidly to the antigen, (c) tissue sections which are morphologically well preserved and in which the antigen has retained its ability to bind the antibody (i.e. has retained its 'antigenicity'), and (d) a means of labelling or otherwise identifying the specifically bound antibody. In practice, the technique has only recently been routinely applied. This is partly because it is only recently that sufficient quantities of antigens of special interest, such as membrane proteins, have become available, and partly because it has proved difficult to obtain satisfactorily fixed tissue sections and at the same time retain the antigenicity of their components. The latter problem is perhaps not surprising, since, when cross-linking and denaturing tissue components during the fixation process, it is to be expected that the molecular features of the antigen for which the antibodies are specific (and to which they must bind) will often be deformed or even destroyed. As

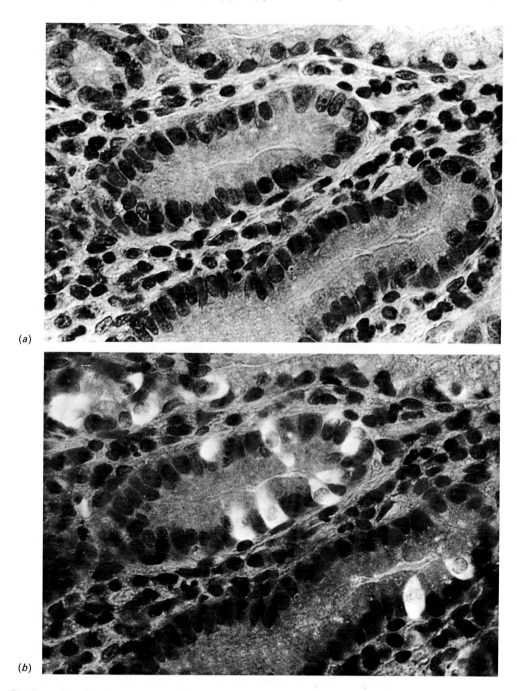

(a)

(b)

Figure 26. *A section through the deeper layer of the stomach wall (gastric mucosa), stained immunocyto-chemically to demonstrate the distribution of the cells containing the hormone, gastrin.*
The section was first treated with an antibody (immunoglobulin) raised in a rabbit against human gastrin. It was then treated with a fluorescein-labelled goat antibody against rabbit immunoglobulin. In (a), the section is photographed with conventional illumination. In (b), the illumination includes light of a wavelength (490 nm) that excites the fluorescein label. The minority population of gastrin-containing cells is brightly fluorescent. Magnification ×1400.
Courtesy of C. Vaillant, Department of Histology and Cell Biology, University of Liverpool.

with enzyme histochemistry, so far the only routine fixatives able to offer a compromise between satisfactory preservation of cell structure and adequate retention of antigenicity are the aldehydes.

Satisfactory methods for labelling antibodies directly have been developed. However, the most effective approach is indirect (Figure 25); it involves raising a 'second' antibody in a second species (for example, the goat) against the antibody which is specific for the tissue component being investigated (raised, for example, in the rabbit). The 'second' antibody is labelled. In this manner, damage and impairment to the binding capacity of the 'first' antibody, which may occur when the label is attached (and upon which everything depends), can be avoided.

The most widely used antibody labels have been fluorescein salts, which can be identified by light microscopy because they fluoresce when excited by light of an appropriate wavelength (see Figure 26). More recently, however, horseradish peroxidase, which, as described above, can be readily identified histochemically in both light and electron microscopes, has often been preferred because its enzymic activity provides a final amplification step.

Virus
Antigen

To date, immunocytochemistry has been employed primarily as a specialized research tool (see Figures 155 and 162b), although it has found some diagnostic value in the detection of viral antigens (e.g. rubella) and immune diseases, in which characteristic antibodies appear in the blood and become concentrated in the glomeruli of the kidney. To exploit the exquisite sensitivity of the technique, however, it is important that its application be carefully controlled. In a rigorous immunocytochemical analysis it is necessary to demonstrate that the labelled antibody is indeed specific only for the antigen being studied.

AUTORADIOGRAPHY

Radioactive isotopes

This is a technical approach that allows the investigator to explore in detail the topographical distribution of radioactive molecules within cells and tissues. It relies upon the knowledge: (a) that as radioactive isotopes decay they emit energized particles, and (b) that these emitted particles affect photographic emulsions in a manner similar to photons of light. The importance of this approach has grown with the increasing availability of radioactive molecules of biological interest to the present time when most 'precursors' can be readily labelled at almost any position within the molecule. As will be seen, optimum resolution of the method is obtained by employing isotopes that emit particles with low energy and thus a restricted path length. For this reason, tritium (^{3}H) (particle energy 0.013 MeV) labelling is to be preferred, although other beta-emitters, with higher particle energies (such as carbon [^{14}C], 0.155 MeV), can also be used.

$MeV = 10^6$ electron volts

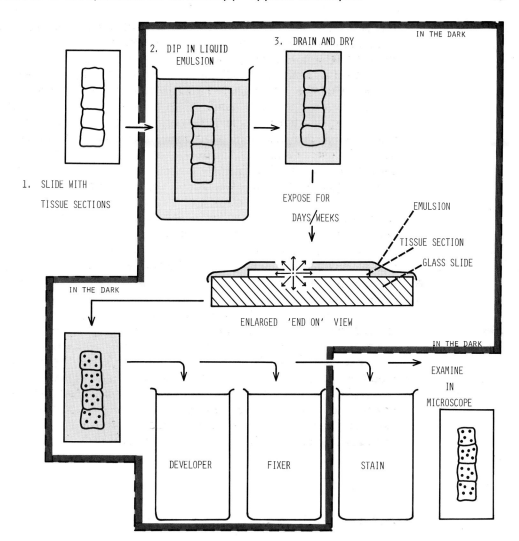

Figure 27. *An outline of the steps involved in preparing sections for autoradiography.*

To examine the intracellular distribution of radioactively labelled material, tissues must be fixed, embedded and sectioned. For light microscopy, wax sections are produced and mounted, as usual, on glass slides; for electron microscope autoradiography, plastic sections are first mounted on grids and then, to make further handling easier, these grids are attached to glass slides and treated as if they were mounted wax sections.

Next, in a darkroom illuminated by light of a wavelength which will not affect the preparation, the upper surface of the sections is coated thinly with a warmed liquid photographic emulsion (Figure 27). The emulsion consists of a gelatin matrix containing evenly distributed silver bromide crystals, and the sections are coated by simply dipping the slides into the liquid and slowly withdrawing them. On cooling and drying, the matrix hardens, adhering tightly to the section surface and holding above it a closely packed, even layer of silver bromide crystals. Slides and grids bearing sections coated in this way are then left in the dark for an 'exposure period' of days or weeks. During this period the isotopically labelled molecules within the tissue section will continue to emit energized particles (in all directions), and a small proportion of these particles will be directed towards the emulsion covering the section surface. Those of low energy, having a restricted path length, will reach only those crystals within the immediate vicinity of the radioactive source. If a particle makes contact with a silver bromide crystal within the emulsion, it causes a physical change within the crystal and a 'latent' image is produced within it. Acquiring a latent image does not produce an overt change in the crystal at this stage, but when it is developed (that is, treated with a reducing solution), it becomes transformed into a crystal of metallic silver. Silver bromide crystals without latent images remain unchanged by the developing solution. The final step in the developing process is accomplished by treating the preparation with sodium thiosulphate (or 'fixer'). This reagent selectively removes the unchanged silver bromide crystals from the emulsion. The metallic silver crystals remain, their position over the isotopic source unchanged. The emulsion is now no longer sensitive to daylight and when stained by routine methods it is ready for examination in the microscope. In the light microscope, silver crystals appear as distinct black dots. With the higher resolution of the electron microscope, the silver grains appear as convoluted, electron-opaque whorls over the labelled component (see Figures 113 and 114). Although electron microscope autoradiography allows the tissue to be examined in more detail, it is worth noting that the autoradiographic record is produced in the same way as for light microscopy; its resolution, in terms of its ability to identify the site at which the isotope is located, is thus very similar.

Other considerations

Most autoradiographic studies follow the synthesis and subsequent fate of a cellular component by inducing the cell to incorporate the labelled precursors instead of native (endogenous) molecules. For these studies it is therefore essential to know something of the structure of the component being synthesized and the conditions under which it is being made. For example, when the synthesis of RNA is followed, uridine (the nucleoside exclusive to RNA) is used in preference to, say, adenosine, since the latter is a constituent of both RNA and DNA. If an exclusive and diagnostic constituent like uridine is not available, it is, instead, often possible to choose conditions which are especially favourable to the manufacture of the component of interest. Thus, for example, the almost

Nucleosides

ubiquitous amino acid, leucine, can be used as a labelled precursor to follow the synthesis of secretory proteins in the acinar cells of the exocrine pancreas, because, when these cells are actively synthesizing their secretion, nine out of every ten of the leucine molecules taken up are used in the synthesis of these proteins. The other, 'structural' proteins of these cells are longer-lived and are made much more slowly.

However, even when a suitable labelled precursor and favourable conditions for its incorporation have been selected, the possibility exists that more than one kind of newly synthesized component may be labelled or that a labelled constituent may be destined to follow more than one kind of intracellular route. For example, when the 'exclusive' precursor, uridine, is used to label RNA, since several different RNA species may be produced, it may follow any or all of several different cellular pathways simultaneously. To a certain extent this difficulty can be identified and perhaps circumvented if, instead of providing the cell or tissue with a diffuse or prolonged supply of labelled precursor, a short, sharp 'pulse' is introduced. This approach is best achieved in practice by using a 'pulse/chase' protocol, in which the radioactive precursor is made available for a brief period, perhaps of less than a minute, before a much larger concentration of unlabelled precursor is added. Under these conditions, a short pulse of radioactive precursor will be taken up by the cell and then 'chased' by unlabelled precursor along the various intracellular pathways. If the labelled molecules enter two separate cellular compartments or areas simultaneously (as, for example, when a pulse of radioactive uridine labels both the nucleus and the mitochondria), it can be concluded that their appearance and what follows is the result of at least two independent events. Using a pulse/chase protocol for a single, related pathway, each cellular compartment should be labelled sequentially (Figure 28).

CELL AND TISSUE CULTURE

For studying the many activities of living cells and tissues that cannot be followed when they occur in the body, methods of maintaining viable cell and tissue elements in artificial conditions in the laboratory have been developed. In practice, these 'in vitro' methods attempt to simulate the 'in vivo' condition by providing a physical and chemical environment in which all of the essential requirements of the cells and tissues are available. *In vitro/In vivo*

In the absence of a vascular supply, cells and tissues in culture must rely solely upon passive diffusion for the exchange of gases, metabolites and waste products, and a ready exchange between the cell surface and the surrounding medium is therefore essential. For organ or tissue culture proper, this is most simply achieved by cutting the tissue into fine, wafer-thin fragments, but, for the prolonged culture of cells, it

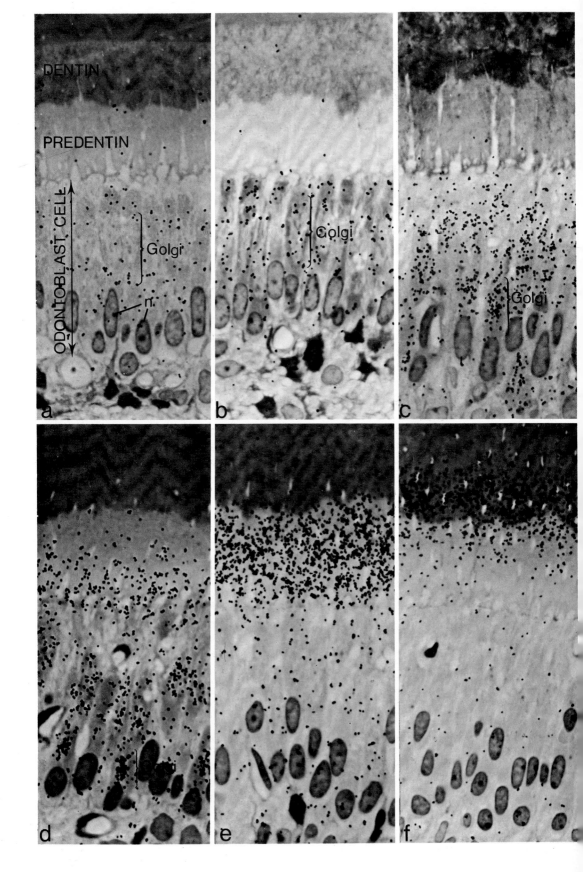

DENTIN

PREDENTIN

ODONTOBLAST CELL

Golgi

n

Golgi

Golgi

a

b

c

d

e

f

is usual to disaggregate the tissue into its constituent cells by incubating it with proteolytic enzymes. The enzymes most often used for dis-aggregation are bacterial collagenases (which digest the connective tissue components that encapsulate and hold most organs together), and trypsin (a protease from the pancreas that digests the adhesive components on cell surfaces). In this way, viable cell suspensions can be produced from most kinds of tissue (see, for example, Figures 128 and 129).

Protease

Culture media and the requirements for growth

The incubation media used to maintain cells and tissues in culture con-tain the essential inorganic ions (i.e. Na^+, K^+, Ca^{++}, Mg^{++}, Cl^-, and PO_4^{--}), glucose, amino acids, and vitamins, together with dissolved oxygen and a buffer which at 37°C will maintain a physiological pH (7.2 to 7.4). Cultured cells can be maintained in such a 'balanced salt solution' for periods of hours, but in the long term and for them to grow and divide it is necessary to add a crude tissue extract derived from either serum or embryonic tissue. It has not been possible, as yet, to identify the essential components of these extracts which are necessary for growth and division, although it is probable that the content of hor-mones (and, in particular, the content of insulin and steroids) is of major importance.

Insulin

Steroids

Recently, a number of specific 'growth factors' that are able to stimulate cells to divide have been isolated and identified. These factors are extremely potent and are able to trigger the division process even in the

---

Figure 28. (*Opposite.*) *A classic example of a study using autoradiography to follow the incorporation, intracellular handling, and subsequent fate of a radioactive precursor. In this instance the amino acid, proline, is being used as a radioactive precursor to follow the synthesis of collagen in the growth region of the tooth.*

Experimental animals were injected with a single dose of the isotope, and then, at intervals, the tissues were removed and prepared for autoradiography. The area of interest – see (a) – shows a layer of polarized, elongated cells (the odontoblasts), a layer of newly secreted tooth matrix (the predentin), and an older layer of mineralized matrix (the dentin).

In (a), at 2 minutes, the silver grains, representing labelled proline (taken up across the base of the cells), are distributed throughout the odontoblast cytoplasm. Note that the Golgi area (see page 161) and the nuclei (n) are not labelled.

In (b), at 10 minutes, the label is beginning to concentrate within the Golgi area, and in (c), at 20 minutes, it is heavily concentrated there.

Since, 20 minutes after the injection, only 20 per cent of the original concentration of label remains in the blood circulation, autoradiographs of preparations taken from this time onwards will largely record the label already taken up. In effect, the tissue thus receives a brief (20-minute) 'pulse' of label.

In (d), at 30 minutes, although there is still label in the Golgi areas, it is now also beginning to be transported to the apical borders of the cells.

In (e), at 4 hours, almost all the labelled proline is within the predentin (the proline is presumed to be incorporated within the protein, collagen, since it accounts for 20 per cent of all amino acids in this molecule).

In (f), at 30 hours, because of the continuing synthesis of (unlabelled) predentin, the cells become displaced basally and the labelled predentin becomes mineralized to form dentin.

All magnifications ×1000.

Courtesy of M. Weinstock and C. P. Leblond, Department of Anatomy, McGill University.

Epidermis

Fibroblast

Chondrocyte

absence of additives like serum. Moreover, their action is specific for certain cell types. Thus far, an epidermal growth factor (EGF), a fibroblast growth factor (FGF), an ovarian growth factor and a chondrocyte growth factor have been identified, but only the molecular structure of the first of these (EGF) has been analysed in any detail (see Figure 77). A nerve growth factor has also been isolated, but its effect on nerve cells in culture is to stimulate growth and branching rather than division (Figure 29).

One useful characteristic that most cells display in culture is that of growing as a thin, pavement-like layer (see page 72), because this 'monolayer' arrangement allows phase or interference contrast optics to be used to examine their form and behaviour directly. Using routine culture methods it is now possible, for example, to grow and examine cardiac muscle cells which will beat rhythmically and macrophages which will actively phagocytose. In vitro systems in which nervous tissue will grow to innervate neighbouring muscle tissue have also been developed.

Dedifferentiation, cell lines and cell ageing

Although in vitro methods are widely used in cell biological research, it is important to realize that, in the long term, cells that adapt successfully to culture conditions usually lose many of their in vivo characteristics. Thus, for example, although pituitary cells and liver parenchyma cells will survive in culture in a differentiated state for several days or even weeks, as a rule they become 'dedifferentiated', losing their characteristic structural organization and their ability to perform 'sophisticated' functions such as the synthesis and secretion of hormones. The most successfully propagated cells in culture are those such as fibroblasts (see page 72), which have a relatively low order of intracellular organization.

Many kinds of cell may, nevertheless, be cultured and will grow and divide to the extent that they can be maintained in vitro over many years. There are now a large number of well-established, so-called 'cell lines' available that have originated in this way. Where they have become fully adapted to the in vitro environment and been shown to be capable of virtually indefinite propagation (i.e. where the number of generations seems unlimited), these preparations have, however, always been found

'Chromosome number' see *Karyotype*

to be in some way abnormal, usually with an altered chromosome number. 'Unaltered', 'normal' preparations in which the cells are, in all respects, apparently identical to their in vivo counterparts can be successfully propagated but it seems that they are able to undergo only

Figure 29. (*Opposite.*) *Nerve cells in culture. These cells were dissociated using trypsin and grown in a medium containing nerve growth factor. The characteristic 'neuronal' form of the cell bodies and their multiple dendritic processes are well shown. Magnification ×40.*
Courtesy of J. M. England, The Wistar Institute, Philadelphia.

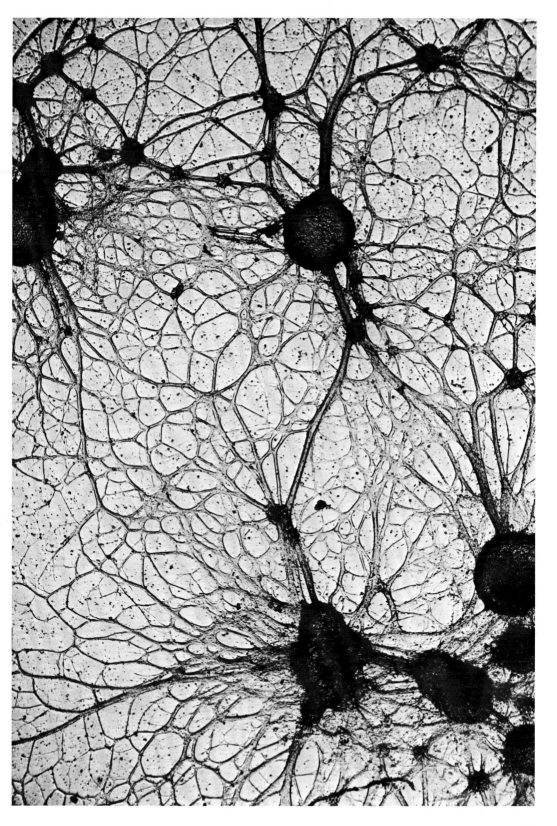

a fixed number of divisions. For human cells a maximum of about 50 divisions is the rule, and the older the donor the more limited is the number of divisions. Because of these characteristics, these preparations have been used as model systems for the study of the ageing process at the cellular level. Cell lines that can be propagated indefinitely are, on the other hand, very convenient to maintain and they are useful for screening

Carcinogen.
Exogenous
chemical agent that
causes cancer.

purposes in laboratories concerned with toxicity or carcinogen testing.

Exceptions to the rule that cultured cells dedifferentiate are found amongst 'immortal' cell lines from tumour cells, in which some residual functional characteristics of the parent cell type sometimes survive. The possibility of fusing this kind of cell with a newly derived differentiated cell type and combining the advantages of both is currently an area of active research interest.

CELL FRACTIONATION

Because they are essentially averaging techniques, the usefulness of most biochemical analyses is directly dependent upon the purity of the preparation being studied. It follows, therefore, that for the satisfactory biochemical categorization of the cellular and subcellular components with which we will be concerned, it is necessary to employ preparative methods, such as those of cell fractionation, which allow any given organelle or class of structure to be analysed free and isolated from the remainder of the cell's components. Although none of the methods of fractionation devised so far provide a cell fraction completely free from contamination by other cell constituents, they have allowed the distinctive biochemical character of most cellular structures to be documented in detail.

In most routine methods of cell fractionation, the tissue is disrupted and its liberated constituents are separated from one another by centrifugation. The method of disruption varies, but it usually breaks down the cellular and intracellular membrane boundaries while keeping discrete organelles such as the nuclei and mitochondria intact. Following their disruption into small fragments, the membrane boundaries, which include the plasma membrane and the endoplasmic reticulum, immediately re-seal to form small spherical vesicles; so the tissue is effectively reduced by disruption to a thin soup containing several classes of membrane-bound particles each with a distinctive size, shape and average density. By layering this preparation over the surface of a suspending medium such as sucrose, and subjecting it to an increased gravitational force by centrifuging, the constituent particles can be separated from each other largely on the basis of their size or their density. Differential centrifugation fractionates primarily on the basis of size, while density gradient centrifugation fractionates largely on the basis of average density.

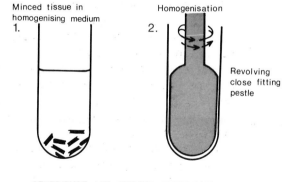

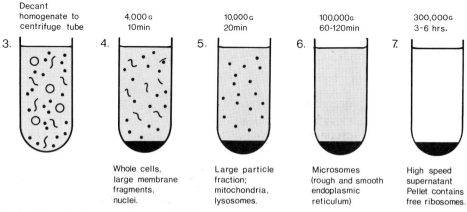

Figure 30. *Cell fractionation by differential centrifugation.*

Differential centrifugation

In this method of fractionation, cellular components are centrifuged through a medium of low density until the largest particles begin to sediment and form a pellet at the bottom of the tube. Once one fraction has pelleted (usually determined by trial and error), the unsedimented remainder, or 'supernatant', can be transferred to a second tube and centrifuged for a longer period and at higher speed to bring a second fraction down. The process is repeated, and, in practice, with a tissue such as the liver, one can obtain successively: an unbroken cell and plasma membrane fraction, a nuclear fraction, a 'mitochondrial' fraction, a microsome fraction, and an unsedimentable 'soluble' fraction (Figure 30). The 'mitochondrial' fraction contains not only mitochondria but also lysosomes and other organelles of a similar size. The microsome fraction is derived largely from endoplasmic reticulum.

Density gradient centrifugation

Particles which fail to separate from each other on the basis of size, such as mitochondria and lysosomes, can often be separated because they have different average densities. In this method, the suspending medium, which is usually sucrose, provides a density gradient within the centrifuge tube. The experimental preparation is layered over the top of the gradient and the tubes are centrifuged at high speed, usually for some hours. During this time the centrifugal force draws the particles into the gradient until they reach a density equivalent to their own average density. Once they have arrived at this point in the gradient their buoyancy prevents them from penetrating further. Usually the purity of the enriched fractions obtained by differential centrifugation can be improved substantially by further, density gradient separation.

Other aspects of separation by centrifugation

The basic principles of centrifugal separation can, of course, be used to separate large elements such as red from white blood cells, and they can also be used under more stringent conditions to separate molecular and macromolecular components such as ribosome subunits (see page 136) and their related ribonucleic acids. In standard conditions, the rate of sedimentation of each of these components is sufficiently predictable for it to be used to categorize the macromolecule involved, and thus a particle or molecule can be given a 'sedimentation coefficient' or 'S' value, measured in Svedberg units. Free liver ribosomes, for example, have a sedimentation coefficient of 80 S, while transfer RNA has a coefficient of 4 S.

PART 2

The nature, form, and function of cell components

The cell theory

The cell theory defined by Schwann and Schleiden in 1838 was the most important biological generalization to emerge during the early years of the nineteenth century. In general terms, the theory declared that all living tissues are composed of cells and cellular products (i.e. if, like connective tissues, they include constituents other than cells then these constituents are of cellular origin). In most respects the cell theory was found to be acceptable to the microscopists of the nineteenth century, although the wide diversity in form of the different cell types stimulated a prolonged controversy over whether it could be applied unreservedly to all tissues. The debate, which lingered on into the early decades of this century, was finally resolved when even the nervous system was shown to consist of cellular elements.

The functional anatomy of the cell

Building upon the foundations laid down by the cytologists of the late nineteenth century, continuing studies in the life sciences have shown the cell to be a discrete, living unit containing all that is required for an independent, self-propagating existence. At the subcellular level the idea (widely held in the early nineteenth century) that the life process depends upon a 'vital' property of an ill-defined 'protoplasm' has given way to the knowledge that cellular processes depend upon a closely knit, self-regulating system of considerable structural complexity. Dissection to the molecular level has shown, however, that within this system there is, nevertheless, a high degree of functional independence and self-control, which in many instances can be related to constituent

Macromolecules.
Include nucleic
acids, proteins,
lipids and
polysaccharides;
large molecules
with molecular
weights between
10^3 and 10^9.

Chromosomes

functional units that have a distinctive molecular architecture. These components may exist as independent assemblies of macromolecules (like chromosomes and ribosomes), or, at a higher level of organization, they may be grouped within membrane lamellae to form organelles (like mitochondria). At this macromolecular level of organization, form and function are essentially the same in all cell types.

A basic plan of the cell can thus be derived which allows us to treat all cells from a common standpoint. As shown in Figure 31, a convenient, though arbitrary, subdivision of the functional anatomy of a typical cell identifies the plasma membrane as delimiting the cell boundary, the nucleus as the store and source of hereditary information, and the cytoplasm as the compartment in which most of the directed, executive functions of the cell are carried out. This subdivision provides the basis of the account that follows.

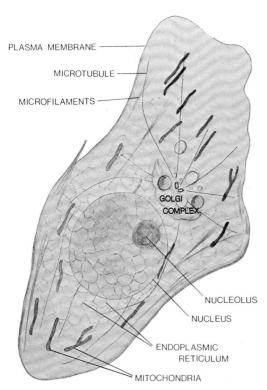

PLASMA MEMBRANE

MICROTUBULE

MICROFILAMENTS

GOLGI
COMPLEX

NUCLEOLUS

NUCLEUS

ENDOPLASMIC
RETICULUM

MITOCHONDRIA

Figure 31. *The essentials of cell structure are best displayed in cells like the fibroblast, which has a relatively low order of intracellular organization. Typically these cells have single central nucleus containing from one to several nucleoli. Within their cytoplasm they contain a variety of membranous elements which (like the endoplasmic reticulum) may either be distributed as an array of tubular or sac-like cisternae, or (like mitochondria) may occur as discrete, individual organelles. The Golgi complex (which is also largely comprised of cisternal elements) is distinctive because it usually occupies a specialized region in the cytoplasm adjacent to the nucleus. This region, which is sometimes called the cell centre or 'cytocentrum', also includes the paired, rod-like particles known as 'centrioles'. Besides these membranous elements the ground matrix of the cytoplasm contains cytoskeletal components (such as microtubules and microfilaments) and free storage deposits (such as lipid and the carbohydrate storage product, glycogen).*
The cell boundary is delimited by a continuous membrane frontier, the plasma membrane.

4

THE PLASMA MEMBRANE

INTRODUCTION

Cellular membranes – their common features

Cellular membranes are discrete, flexible lamellae that envelop the cell and enclose its cytoplasmic compartments. They always provide closed, continuous boundaries.

Lamella. A thin flat sheet.

Chemical analyses show that all eukaryotic cellular membranes consist primarily of protein and lipid, and, although each type of membrane clearly has special features related to its function, their common physical characteristics (such as their surface tension, buoyant density, electrical resistance and water permeability) indicate that they are all essentially similar structures. When sectioned, stained appropriately, and examined in the electron microscope, eukaryotic cellular membranes always show a three-layered substructure of two dense outer leaflets separated by a less dense central layer (Figures 32 and 33). The consistency of this appearance again reflects the basic similarity of the molecular organization in all cellular membranes.

Eukaryote

The plasma membrane

All cells are enveloped by a 'plasma membrane' that is usually about 7.5 nm thick. This membrane provides the boundary that mediates all the interactions between the cell and its environment. In carrying out this function it acts primarily as a selective barrier concerned with the inward transport of metabolites and the outward transport of waste products. However, since it is able to respond to changes in the extracellular environment, by generating signals that influence intracellular events, the plasma membrane also plays an important role as a receiver and transducer of information.

In mobile cells the plasma membrane is also involved in the process of locomotion, whereas in solid tissues it is responsible for intercellular adhesion.

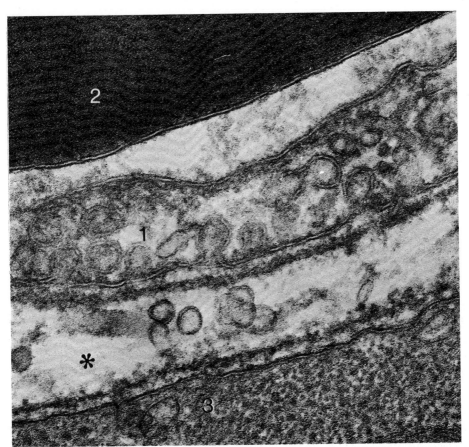

Figure 32. *An electron micrograph of a section across the wall of a blood capillary. The endothelial cell that forms the capillary wall runs across the middle of the field (1). In the capillary lumen, the edge of a red blood cell is seen (2), while outside, separated by an intercellular space (*), is the edge of a muscle cell (3). The section plane runs normal to the surface of all these cells and clearly shows the trilaminar substructure of their plasma membranes.*
  Courtesy of N. and M. Simionescu, Section for Cell Biology, Yale University.

PLASMA MEMBRANE STRUCTURE

Recent studies on the plasma membrane have derived great benefit from the introduction of two new techniques. Before dealing with the current views on its structure, it will therefore be useful to learn something of each of them.

Freeze-fracture

The first approach is a method of tissue preparation known as freeze-fracture, and the main steps are outlined in Figure 34. To begin with,

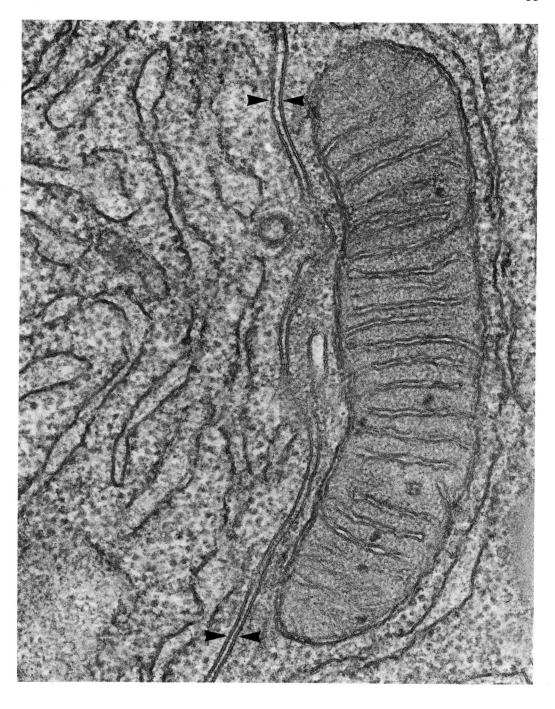

Figure 33. *An electron micrograph showing the periphery of two cells in the exocrine pancreas. The trilaminar appearance of the plasma membranes (arrows) and of the membranes of the intracellular organelles is clearly evident. Magnification ×88 000.*

Courtesy of N. and M. Simionescu, Section for Cell Biology, Yale University.

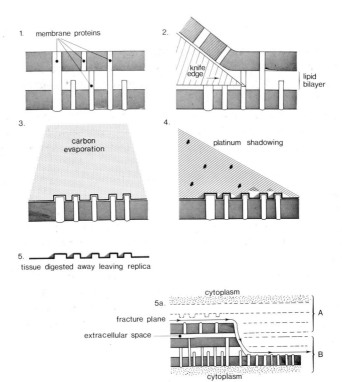

Figure 34. *A schematic outline of the freeze-fracture technique (for details refer to the text). Diagram 5a should be viewed with reference to Figure 39. When a fracture plane crosses the boundary between two cells (A and B), it causes a step-like discontinuity and exposes different aspects of the inner membrane in the two cells. In cell A, the face with the extracellular space behind it (the extracellular or 'E' face) is seen; in cell B, the other face (the one with the cytoplasm behind it — the protoplasmic or 'P' face) is exposed. Typically, the P face has a denser population of intramembranous particles than the E face.*

a suspension of cells or a small block of tissue is frozen rapidly in liquid nitrogen (−180°C) and then fractured under vacuum. The fracture plane passes through the interior of the plasma membrane and exposes large areas in a three-dimensional relief. For transmission electron microscopy, a thin, precisely faithful replica of the fractured surface topography is made by evaporating carbon (heated in a vacuum) as a thin film onto the exposed surface. The protuberances and hollows of this carbon replica are then emphasized by evaporating platinum (again heating in a vacuum) and casting it, as a fine drift-like layer, at an angle across its surface. Finally, the tissue is digested away from beneath and a thin carbon—platinum film remains. A three-dimensional replica, thin enough to be examined in the transmission electron microscope, is thus obtained. The advantages of this technique are: (a) it allows the membrane to be studied without chemical fixation or staining, (b) it provides relatively large areas of membrane for examination, and (c) as described below, it offers a new aspect, the membrane internum, for study.

Lectins

Lectin. From the Latin, 'to select'.

The second kind of technical approach is directed at the extracellular surface of the plasma membrane and is based upon the knowledge that a number of proteins and glycoproteins extracted from plant and animal tissues can bind selectively to specific sugar residues. The most widely used of these agents (which are known generally as 'lectins') have been Concanavalin A (from the jack bean *Canavalia ensiformis*), which binds to glucosyl residues, and wheat germ agglutinin, which binds to *N*-acetyl-*D*-glucosamine.

Glycoproteins. Large protein–carbohydrate molecules in which chains of sugar residues are covalently linked to the polypeptide 'backbone'.

Glycoproteins bearing these sugar residues are important functional components of the cell surface and, by using lectins to which electron-opaque markers or radioactive labels have been attached, they can be readily identified. One particular advantage of this approach is that it allows glycoproteins to be mapped and measured on the surfaces of both living and fixed cells.

Another important feature of most lectin molecules is that they can each bind two or more sugar residues. This means that they are able to act as a bridge between adjacent sugars. As will be described below, the ability of lectins to cross-link cell surface sugars has proved a useful means of manipulating cell membrane components.

The arrangement of lipids and proteins in the plasma membrane

One of the earliest attempts at describing a generalized membrane sub-structure is the model originally proposed by Gorter and Grendel and later defined more precisely by Danielli and Davson as the 'lipid bilayer model'. On the basis of the available data (morphological, chemical and physiological), this model still provides the most widely accepted account of the distribution of lipid molecules in membranes. Its major feature is a central, liquid bilayer of lipid molecules that forms the back-bone of the membrane and serves as its primary permeability barrier. Most membrane lipid molecules have hydrophilic (polar) and hydro-phobic (non-polar) ends (see Figure 35) and lie within the membrane with their hydrated polar groups facing outwards and their disordered hydrocarbon chains (non-polar ends) directed inwards (Figure 36).

It is believed that, while the lipid molecules within either the upper half or lower half of the bilayer can easily exchange places with their laterally adjacent neighbours (i.e. those in the same monolayer), an exchange of lipid molecules across the bilayer, from one side to the other, rarely occurs.

In the original Danielli–Davson model, the lipid bilayer was sand-wiched between continuous layers of 'unrolled' protein, but it now seems more likely that most membrane proteins are globular and have an irregular, discontinuous distribution. Indeed, in the most widely accepted contemporary model of membrane structure (the 'fluid mosaic

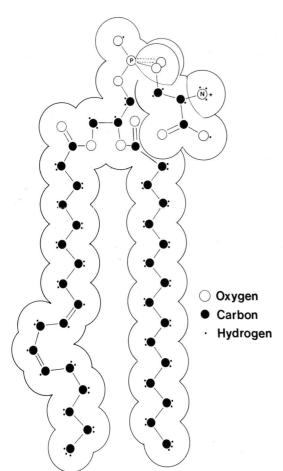

Figure 35. *A phospholipid molecule. In membranes this is the most abundant kind of lipid molecule. It consists of a three-carbon framework attached to a polar group and two long-chain fatty acids. The phosphate is contained within the polar group.*

○ **Oxygen**

● **Carbon**

· **Hydrogen**

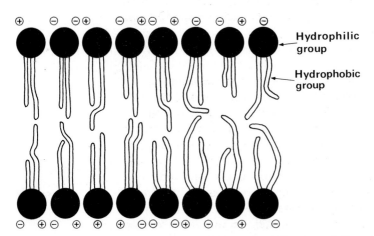

Hydrophilic group

Hydrophobic group

Figure 36. *The lipid bilayer, with the charged hydrophilic groups facing outwards.*

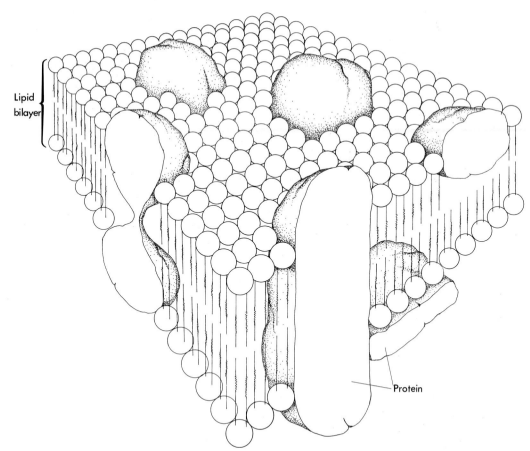

Lipid bilayer

Protein

Figure 37. *The fluid mosaic model. After S. J. Singer and G. L. Nicolson.*

model' of Singer and Nicolson) it is proposed that most integral membrane proteins are independent and quite separate entities that float upon and within the liquid phase of the lipid (Figure 37). Membrane proteins that are associated with the membrane surface, but do not even partially penetrate the lipid bilayer, are regarded as 'peripheral' rather than 'integral' components.

The Singer–Nicolson model has received direct support from the few detailed studies that have so far been made on the structure and disposition of integral membrane proteins. These studies have shown that integral membrane proteins may be linearly 'amphipathic', which means that along their length they have a series of amino acid sequences that are predominantly hydrophobic and a series of sequences (most notably at their amino terminal ends) that are predominantly hydrophilic. It appears that these amphipathic components lie with their hydrophobic regions embedded within the membrane internum (probably within the hydrophobic lipid layers), while their hydrophilic regions are exposed to the hydrophilic environment at the membrane surfaces.

Amino acids

'Hydrophobic/
Hydrophilic' see
Polar molecules

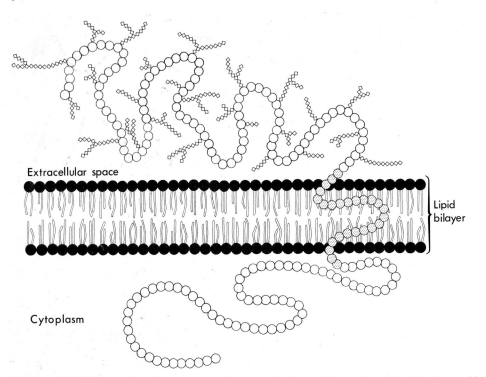

Extracellular space

Lipid bilayer

Cytoplasm

Figure 38. *The amino terminal end of the glycophorin molecule extends over the external surface and bears perhaps as many as 30 branching side-chains of sugar residues (◇). Because of these sugar residues, this external portion (consisting of about 100 amino acid residues) is hydrophilic. It continues through the lipid bilayer as a region of about 25 residues (○) which includes only uncharged (hydrophobic) amino acids like serine, proline and leucine. On the cytoplasmic surface of the membrane the molecule terminates at its carboxyl end as a linear chain of about 60 amino acid residues, and the majority of these are charged, i.e. they are hydrophilic. After V. T. Marchesi.*

The most detailed evidence so far available on the amphipathic nature of integral membrane proteins has been obtained from the structural analysis of the major glycoprotein of the erythrocyte membrane. These studies have shown that this molecule (known as glycophorin — mol. wt. about 50 000) is comprised of a single chain of 200 amino acid residues. The first 100 residues (at the amino terminal end) bear many side branches rich in carbohydrate sugars, and these can be shown to be exposed on the extracellular surface of the membrane (Figure 38). The middle 25 or so amino acids are uncharged (i.e. hydrophobic), while the remainder (about 60 residues) are charged and probably exposed on the inner, cytoplasmic surface. The glycophorin molecule is thus linearly amphipathic and conforms in all respects with the thermo-dynamic requirements of a membrane protein that penetrates and crosses the hydrophobic lipid bilayer.

Good evidence for the presence of protein components within the lipid bilayer of the plasma membrane is also available from freeze-fracture studies. In this technique, the fracture plane in the cleaved tissue actually travels through the lipid internum of the cell membranes and splits them into two complementary sheets (see Figure 34). The inner

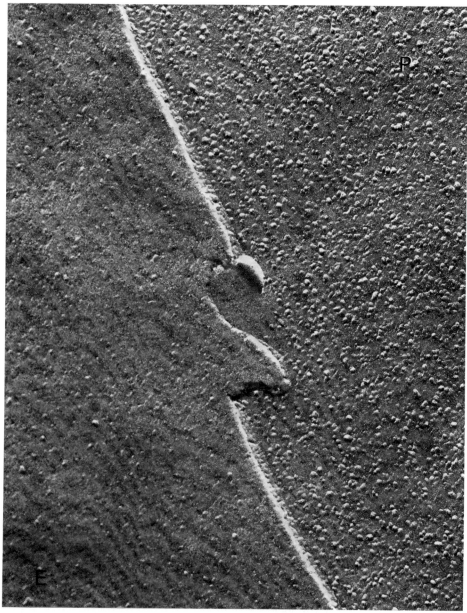

Figure 39. *Freeze-fractured plasma membranes from two neighbouring cells in the endocrine pancreas. For orientation, refer to Figure 34. On the left-hand side, the E face of one cell's membrane is displayed, while on the right-hand side is the P face of the other cell's membrane. The P face clearly has a denser population of intramembranous particles.*

    Courtesy of A. Perrelet and L. Orci, University of Geneva.

aspect of membranes thus exposed consistently reveals the presence of intramembranous particles 6 to 12 nm in diameter (Figure 39). When the exposed surface glycoproteins of erythrocytes are induced to aggregate into clusters on the cell surface by lectins (see below), it has been found that the intramembranous particles also aggregate to form the same clustered pattern. These and other observations indicate clearly that intramembranous particles represent groups of transmembrane proteins that either partially or completely penetrate the lipid bilayer.

Plasma membrane carbohydrates

Earlier microscopical studies consistently identified an acidic glycoprotein layer, often called the 'glycocalyx', on the extracellular surface of eukaryotic cells. Rather than being an 'extraneous coat', this layer is more properly considered an integral part of the plasma membrane, since many, if not most, of its constituent glycoproteins probably have their non-polar regions embedded within the central lipid bilayer. The exposed polar ends of these molecules bear carbohydrate side-chains rich in neutral sugars which usually end with a terminal fucose or sialic acid residue.

The distribution of the sugar-bearing ends of the membrane glycoproteins on the external surface is an important feature related to the functional asymmetry of the plasma membrane, since many of these sugar-bearing termini provide the basis of much of the discriminating specificity that is one of the major responsibilities of the external cell surface (see below). Few, if any, carbohydrate residues belonging to plasma membrane glycoproteins are exposed on the inner, cytoplasmic surface of the membrane.

The plasticity of the plasma membrane

Molecular models and electron micrographs of thin sections and freeze-cleaved replicas tend to convey the impression that the plasma membrane is a rigid, immobile boundary. However, while it is true that in many tissue cells the plasma membrane has a characteristic and highly differentiated form (for example, columnar epithelial cells – see Figure 51), it is probable that, in the plane of the membrane, many molecular constituents are continuously in motion. This idea is derived from theoretical considerations which predict that (due to Brownian motion) the fluidity of the lipid bilayer will allow a continuous state of interchange between neighbouring membrane proteins. However, it has also been shown directly in experiments in which membrane proteins have been induced to move rapidly (and in concert) over the cell surface. In one of the most convincing demonstrations of membrane fluidity, the ability of a surface-active virus to fuse cell membranes (see page 127) was exploited in order to fuse cultured human cells with cultured mouse cells. Species-specific (i.e. mouse or human) membrane proteins were then identified on the hybrid cell surface using immunofluorescence and it was found that, although these components were initially confined to

Brownian motion

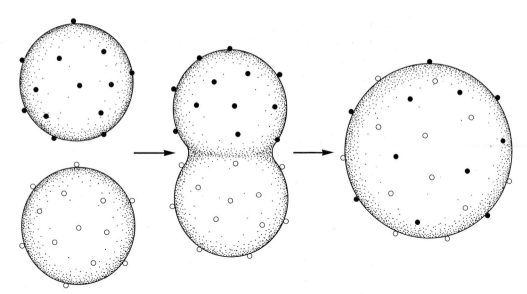

Figure 40. *The fusion of cultured mouse and human cells using Sendai virus. Species-specific components on the surface of the hybrid cells are initially restricted to their respective hemispheres but within minutes they intermix. From the work of Frye and Edidin.*

their respective halves, they soon intermixed (i.e. within minutes) and became evenly distributed over the whole surface (Figure 40).

   In some cells, lectins like Concanavalin A, which can bind and cross-link several neighbouring sugar-bearing proteins (see above), have been shown to be able to induce plasma membrane components to move in the plane of the membrane. Since the lectins can be labelled with fluorescein, their movement can be followed over the cell surface, and, most typically, as shown in Figure 41, they show a definite pattern of redistribution that eventually leads to the formation of a cap at one pole of the cell. Only cross-linked components move (unbound components remain undisturbed) and, while it is known that their movement is dependent upon both their cross-linking (lectins unable to cross-link are unable to induce capping) and the fluidity of the membrane lipid (it is, for example, prevented by low temperature), the agencies responsible for driving and directing it are unknown.

Membrane fluidity and cell locomotion

When mobile cells such as granular leucocytes and cultured fibroblasts move over a flat surface they display a distinctive pattern of shape changes. As a rule, locomotion begins with the extension of a leading edge of the cell as a broad, flowing pseudopod (the ruffled border) that becomes attached on its lower surface to the substratum. Once attachment is secured, the main body of the cell is then drawn forwards,

Leucocytes

Fibroblasts

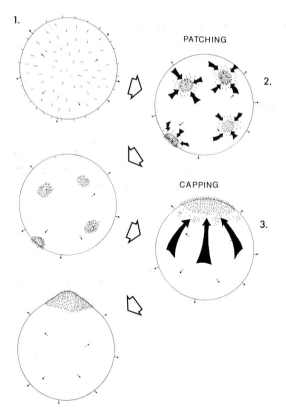

Figure 41. *Patching and capping of plasma membrane components induced by a lectin. In (1), the even distribution of two different components (| and ↑) on the free cell surface is shown. In (2), the addition of a lectin able to specifically bind and cross-link one of the components causes the component to become aggregated into patches. In (3), the patches coalesce to form a polar cap; the unbound components retain their original distribution. By using labelled lectin, the progress of these events can be followed in living cells.*

towards the leading edge. On the upper surface of the extending pseudopod, just behind the leading edge, the plasma membrane of the moving cell is usually thrown into a series of folds (hence the name 'ruffled border') and it has been shown that, even as the leading edge is extending forward, the membrane forming the ruffles on this surface is in fact moving backwards, towards the cell body. Currently, our understanding of the molecular events that underlie these extensive redistributions of the cell membrane is poor (for further consideration, see page 227), but clearly here, as in lectin capping, membrane components are moving independently of their immediate surroundings.

All of these observations on the fluidity of the plasma membrane are entirely consistent with a structure in which the protein-containing components are seen as 'icebergs' moving on and in a 'sea' of liquid lipid.

This analogy, which is frequently applied to the fluid-mosaic model, is worth remembering when the fixed, static profiles of electron micrographs are examined.

Peripheral membrane proteins

Essentially unmovable protein components also exist in the plasma membrane and it appears that the membrane components of some cell types are much less mobile than those of others. Membrane proteins are probably restrained in one of two ways, one deriving from below the membrane as a system of multiple anchors and the other determined by the distribution of junctional elements in the plane of the membrane (discussed below). Anchoring components that have been identified on the cytoplasmic surface of plasma membranes and have been shown to restrict the movement of integral membrane components include the protein, spectrin, a major component of the inner surface of the human erythrocyte membrane. Other components are less well defined but

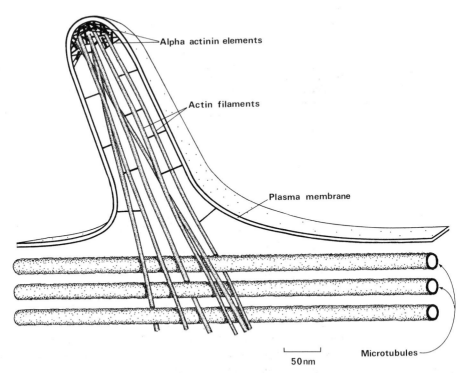

50 nm

Figure 42. *A reconstruction intended to provide an idea of the relative dimensions of the plasma membrane and its associated cytoskeletal elements. The overall thickness of the plasma membrane is about 7.5 nm. The diameter of the actin filaments is between 5 and 8 nm; they are shown projecting into a short microvillus, where they associate with the plasma membrane via bridging elements of alpha actinin (rod-like molecules 30 nm long, 2 nm diameter). Running parallel to the cell surface are the microtubules (outer diameter about 24 nm). After F. Loor.*

probably include those proteins such as alpha actinin and desmin that are believed to be responsible for the insertion and attachment of contractile proteins to the inner surface of the plasma membrane (Figure 42) (for further discussion, see page 232). Components concerned with the attachment of cytoskeletal elements such as microtubules almost certainly also occur, but remain to be identified.

Cytoskeleton

PLASMA MEMBRANE FUNCTION

THE ROLE OF THE CELL SURFACE IN RECEIVING EXTRACELLULAR SIGNALS

For the tissue and organ systems of the body to be functionally integrated their constituent cells must be able to communicate with one another. They do so using messenger molecules such as hormones and neurotransmitters that are released into the blood and tissue fluids by the cells that synthesize them. For a hormone to influence intracellular events within a 'target cell', it must either penetrate the plasma membrane or be able to transmit a signal across it. Steroid hormones, such as oestrogen and progesterone, probably because they are lipid soluble and thus able to penetrate the central bilayer of the membrane easily, readily cross the plasma membrane and enter the cell. Glycoprotein hormones, on the other hand (e.g. the pituitary gonadotrophins that act on the gonads), bind to the surface of their target cells and, without penetrating the plasma membrane, exert their influence from there.

Steroids

Gonadotrophins

The stimulation of a target cell by most non-steroid hormones therefore depends upon the presence of receptors on the outer cell surface. These receptors are integral plasma membrane components and can be identified by their ability to bind specifically to a particular hormone molecule (Figure 43). A target cell that is stimulated by a variety of different hormones (for example, fat cells respond to at least nine different hormones) will bear several populations of surface receptors, each specific for one hormone. As a general rule, there are about 10 000 receptor molecules for each hormone on each cell, and stimulation occurs when a sufficient number of them become occupied. The mechanisms concerned with transducing the initial signal and eliciting the required cellular response (which, for example, in a fat cell stimulated by the hormone, insulin, is lipid synthesis) remain to be fully elucidated. In many cases, however, it is thought that cyclic nucleotides, such as cyclic adenosine monophosphate (cyclic AMP), which are located within the plasma membrane probably play the most important role of 'second messenger'. The binding between the hormone and its receptor is concentration-dependent and appears to be reversible, so that, when the concentration of hormone in the environment around the target cell decreases, the bound hormone molecules probably dissociate from their receptors.

Cyclic AMP

Figure 43. *A transmission electron micrograph showing the distribution of hormone receptor sites on the surface of a pituitary cell. The hormone has been linked to ferritin, an iron-containing tracer that appears here as small, intensely electron-opaque particles. The tracer becomes associated with the cell surface because the hormone it carries binds specifically to its receptor sites. The distribution of the tracer (arrows) thus indicates the distribution of the receptors. Magnification ×90 000.*

Courtesy of C. R. Hopkins and H. Gregory, Department of Histology and Cell Biology (Medical), University of Liverpool.

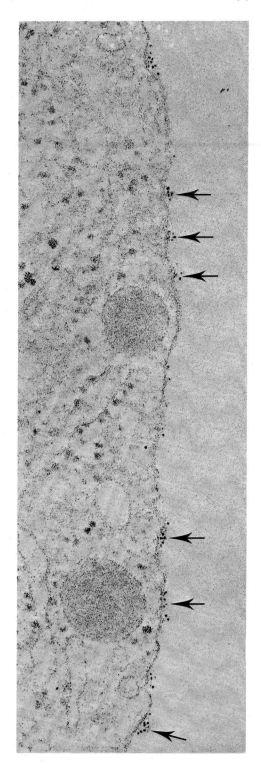

The presence of receptor molecules on the extracellular surface of the plasma membrane in target cells is the basis of many kinds of cell communication. For example, neurotransmitters released from nerve endings (e.g. acetylcholine and noradrenaline) act in a similar way to protein hormones; lymphocytes, the blood- and lymph-borne cells concerned with general surveillance in the body's defence system, also depend upon the ability of their surface receptors to detect and bind invading 'foreign' molecules (antigens). The chemical identity of these receptor molecules is, in most instances, not yet known, but it is probable that they, too, are integral glycoproteins.

THE PLASMA MEMBRANE AND MALIGNANCY

Virus

Malignant tumours arise in the body spontaneously but they may also be induced by radiation, chemical agents and virus infection. The most fundamental abnormality of malignant cells is that they are neoplastic, i.e. they divide without stimulus and are not subject to the normal influences of growth control (see page 111). As malignant tumour cells proliferate they generally display invasive behaviour, penetrating the surrounding tissues. When they reach circulatory systems or body cavities, groups of cells may detach from the tumour and become transported throughout the body. This process of cell dissemination is known as metastasis and it leads to the formation of multiple secondary growths. Both invasive behaviour and the ability to metastasize are probably related to special properties of the malignant cell surface.

Chemical agents and viruses that are able to produce tumours in vivo are also able to alter ('transform') cells in culture. When cells 'transformed' in this way are injected into experimental animals they, too, are usually able to produce tumours. Since cells in culture can be readily transformed to produce homogeneous preparations that are easily maintained, they are widely used for studying malignant change in preference to the more heterogeneous preparations derived from tumours.

Amongst the most prominent differences between transformed cells and their untransformed counterparts are those related to the cell surface.

Contact inhibition

Fibroblasts

One aspect of the behaviour of transformed cells that may be related to their invasive ability is their distinctive pattern of growth in culture. For example, in optimum culture conditions a sparse population of untransformed fibroblasts divides repeatedly until it produces a confluent pavement-like monolayer of cells. The cells become evenly spread to form the monolayer for two reasons: (a) because they are actively mobile (at least when not dividing), and (b) because they are unable to crawl over each other. This means that, as the population density increases and contact becomes more frequent, movement is increasingly restricted. Eventually, when the cells reach confluence, all movement (and division) ceases and the population as a whole is then said to be 'contact inhibited'.

It is a characteristic feature of transformed cells that they are not subject to contact inhibition and thus, when they reach confluence, their continued division eventually leads to an extensive overlapping of cell boundaries (Figure 44). The reason transformed cells are no longer liable to contact inhibition is probably because their cell surface is altered, but it is not known if the alteration is related to their failure to receive rather than send the contact inhibiting signal. In this context, it is of interest that a clearly identifiable feature of 'normal' cells, which is typically absent from transformed cells, is a large plasma membrane glycoprotein (mol. wt. about 250 000) known to be exposed on their external surface (the so-called LETSP or 'large exposed transformation sensitive protein'). On the surface of contact-inhibited cells the amounts of this glycoprotein are increased.

The available information concerning the adhesive properties of the transformed cell surface is very incomplete and cannot yet be related to the in vivo ability of malignant cells to metastasize. There does, however, seem to be a generalized alteration in the surface adhesiveness of transformed cells, because they behave differently with surface-active

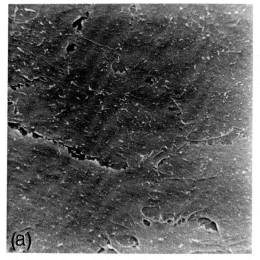 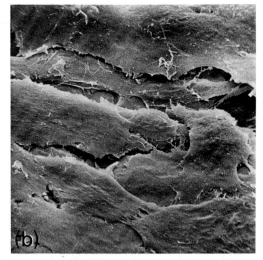

Figure 44. *Scanning electron micrographs of (a) normal and (b) transformed cells in culture. In both preparations the cells have multiplied until they formed a confluent monolayer. In the normal cells, further growth and division is then arrested (i.e. the cells are 'contact inhibited') but in the transformed cells, proliferation continues and the cells overgrow their neighbours. Magnification ×1000.*
   Courtesy of J. G. Collard and J. H. M. Temnink, the Netherlands Cancer Institute.

agents like lectins. When incubated in suspension with wheat germ agglutinin, for example, transformed cells always clump together and aggregate much more readily than their untransformed counterparts. As mentioned above, it is known that aggregation occurs because each lectin molecule is able to bind sugars at more than one site, and it can thus act as a bridge between the sugar-bearing surfaces of neighbouring cells. However, it is not easy to account for the increased agglutinability of transformed cells, because it has been shown that they contain the same density of lectin binding sites as those of 'normal' cells.

Protease

A positive rather than a negative feature of the transformed cell surface is its possession of a specific neutral protease (plasminogen activator) that is able to activate other serum-borne proteases. The functional significance of this protease is unknown but it is of interest that untransformed cells agglutinate much more readily if they, too, have first been treated with a protease.

THE PLASMA MEMBRANE AND TRANSPORT

Homeostasis, the maintenance of the intracellular environment, requires controlled, selective transport across the plasma membrane. The transport mechanisms that exist in the plasma membrane include both microtransfer and macrotransfer processes.

Microtransfer processes

Microtransfer includes the transport of ions and small molecular species such as amino acids and simple sugars; it is carried out by both passive and active mechanisms:

Passive and facilitated transport

When uncharged substances (i.e. non-electrolytes) move across the plasma membrane, they tend to flow down the concentration gradient at a rate determined by their relative concentration on each side of the membrane ('diffusion'). For charged substances, however, the electrical concentration needs to be considered in addition to the concentration gradient, because if charged substances move across a selectively permeable barrier like the plasma membrane they will also generate a potential difference. This aspect will be discussed further below, but it is worth noting here that a continuous membrane backbone composed of a bilayer of lipid molecules with their charged groups facing outwards presents a formidable barrier to the passage of both charged and lipid-insoluble substances. Because of this, and because of what is known

of the transport of different ions across the plasma membrane, it has *Ion*
been predicted that hydrophilic channels or pores allowing the transport
of water and ions through the plasma membrane must exist. However,
although fairly precise estimates of the size and distribution of these
pores have been made (in the erythrocyte, for example, they have been
calculated to be between 0.7 and 0.8 nm in diameter and are believed
to occupy about 0.06 per cent of the total surface area), they have not
been identified morphologically.

Certain molecules, such as glucose and glycerol, cross the plasma
membrane many times faster than can be accounted for by passive diffusion alone. Their transport is thus said to be 'facilitated' or 'catalysed'
and it is presumed to be regulated by specific catalytic components in
the membrane. These components are probably protein in nature and,
like enzymes, although they do not change the final concentration at
equilibrium, they do increase the rate at which equilibrium is reached.

Active transport

Also present in the plasma membrane are transport systems that require
energy expenditure by the cell. These systems can operate against the
concentration gradient and are thus able to achieve a several-fold
increase in concentration on one side of the membrane under physico-
chemical conditions that would otherwise result in an equal concentration on both sides. The best-documented active transport mechanism
carries sodium ions (Na^+) out of the erythrocyte and accumulates
potassium ions (K^+) within it. In this coupled pumping mechanism,
a membrane-bound magnesium-activated adenosine triphosphatase
(ATPase) hydrolyses one adenosine triphosphate (ATP) molecule (to *Adenosine*
adenosine diphosphate or ADP) to provide energy for every three Na^+ *triphosphate*
and two K^+ ions transported.

Most models of active transport postulate that 'carriers' translocate
the ion (or amino acid, sugar, etc.) from one membrane surface to the
other and in so doing lose their energy. It is presumed that the carrier
components can then be regenerated by gaining energy from some
energy-yielding reaction involving adenosine triphosphate. From
thermodynamic considerations alone it is, however, very improbable that
any component can move with facility from the hydrophilic environment
of the membrane surface back and forth across its hydrophobic internum.
It is now regarded more likely, therefore, that 'carriers' for active transport
are integral transmembrane components whose ability to transport varies
with their energy state, but whose topographical position within the
membrane remains unchanged. Recently, the ATPase responsible for
Na^+ transport across the plasma membrane in the kidney (see below)
has been isolated and identified; in keeping with the idea of a 'fixed
carrier' component, it has been shown that one end of this molecule
is exposed to the cytoplasm, and the other, the sugar-bearing end, is
exposed on the external surface.

The distribution of the major monovalent ions across the plasma membrane

The interrelationship between the passive diffusion and the active transport of ions across the plasma membrane is complex, and it is different in different cell types. In general, however, it can be said that an electrical potential exists across the plasma membrane in all cells, and this potential is primarily due to the uneven distribution of charge-carrying ions.

Although the distribution of any particular ion across a plasma membrane is determined partly by its own concentration, its distribution is also profoundly influenced by the other charges in the system. The overall distribution is defined by the Gibbs–Donnan equilibrium, which predicts in electrochemical terms that a system, such as that in existence across the boundary of a cell, will only be in equilibrium when the sum total of the positive and negative charges on either side of the membrane are equal. This means that charged species can move across the membrane only when their transfer is balanced by a movement, in the opposite direction, of ions carrying an equal and similar charge. Thus, although in circumstances in which there is an imbalance in concentration across a membrane there may be a tendency for an ion species to flow down its concentration gradient, its transfer will not be allowed to disturb the electrochemical equilibrium.

In considering the total ionic balance across the plasma membrane, the variable permeability of the membrane to different ions is also important. Thus, although the plasma membrane is normally freely permeable to water, and is probably highly permeable to anions like Cl^- and HCO_3^-, it is impermeable to many negatively charged cytoplasmic components. In some cell types these impermeant anions (which may include amino acids, molecules like ATP, and complex ions like isethionate) make up as much as half of the total intracellular negative charge.

Between these extremes of free permeability and impermeability, the plasma membrane also displays a selective permeability towards some ions; in respect of the major monovalent cations in particular, the membrane permeability to K^+ is usually as much as fifty-fold greater than its permeability to Na^+ (the differential permeability in this instance is probably due to the larger radius of the hydrated sodium ion).

Taking all of these considerations into account, and with the knowledge that in the plasma membrane of many, if not most, cells the major active transport system is concerned with pumping Na^+ outwards and K^+ inwards, a generalized scheme for the relative distribution of the major monovalent ions across plasma membranes can be outlined, as shown in Figure 45.

The magnitude of the potential difference that exists across the cell membrane varies from cell type to cell type, but the overall charge of the cytoplasm is always negative relative to the external environment. It is primarily dependent upon the size and direction of the K^+ ion gradient. Some of the absolute values that have been obtained for the

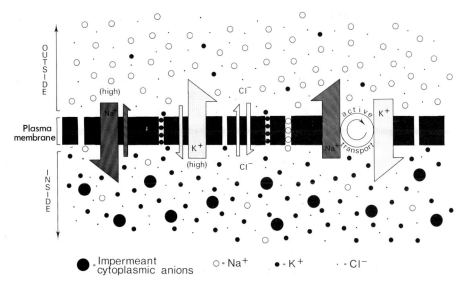

Figure 45. *A schematic representation of the distribution of monovalent ions across the plasma membrane. The size of the channels allowing passive diffusion across the membrane indicates the membrane's selectivity for the various ion species; the size of the arrows indicates the magnitude of the different transport processes. In the plasma membrane of a nerve, it has been estimated that there may be as few as two Na^+ channels/μm^2.*

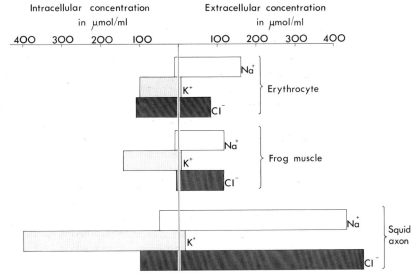

Figure 46. *The distribution of the major monovalent cations across the plasma membrane in the tissues in which it has been measured directly.*

distribution of the major monovalent ions across the plasma membrane are given in Figure 46.

In the erythrocyte, the non-diffusible (impermeant) intracellular anions include haemoglobin and ATP. These anionic components make a rela-   *Haemoglobin*
tively small contribution to the total internal charge, and so for the imbalance in ionic concentration (and the resultant potential difference)

across the plasma membrane to be maintained, it is essential for both Na^+ and K^+ to be actively transported continuously. As mentioned above, in the erythrocyte the active export of Na^+ and import of K^+ is indeed coupled.

In nerve and muscle cells the non-diffusible anions provide a more significant contribution to the intracellular negative charge, and in these cells the active transport of Na^+ alone should be sufficient, theoretically, to maintain the potential difference that exists across the plasma membrane. In fact, however, there is evidence in both of these cell types that K^+ ions are also actively transported, although the extent to which their transport is coupled to Na^+ is not clear.

Ion transport and cell volume

Because water permeates through the plasma membrane readily, its transport into the cell is determined largely by osmotic forces, and these, in turn, depend upon the relative concentrations of ions in the internal and extracellular environments. Control of ionic balance thus allows the direct control of cell-water content, and it is this that dictates cell volume.

Excitable cell membranes

The action potential

An intrinsic property of the plasma membranes of neurones and striated muscle cells is that when they become depolarized (i.e. when the potential difference across the membrane is sufficiently reduced) their permeability to Na^+ increases. Because the potential on the cytoplasmic side of the membrane is negative, there is then an inward movement of Na^+ ions, and this causes the membrane to depolarize further (see Figure 47). The Na^+ permeability is therefore then further increased. This amplifying cascade continues until eventually (within a millisecond of being stimulated) the potential difference across the plasma membrane becomes first abolished and then reversed. This change in transmembrane potential is known as an 'action potential'. In its latter stages, conditions revert rapidly to the original, 'resting potential', primarily because an increase in the permeability of the membrane to K^+ (0.5 to 0.6 milliseconds) follows the initial Na^+ influx (Figure 47).

Propagation of the impulse

Neurone

1. *The local circuit effect.* The propagation of a wave of excitation along the membrane of an elongated cell, like a neurone or a muscle cell, may be brought about by the generation of a local circuit, as shown in Figure

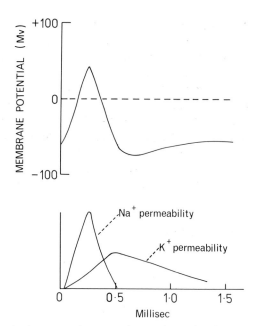

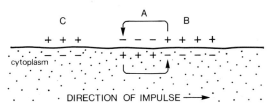

Figure 47. *An action potential (above) shown on the same time scale as the changes in relative permeability to Na⁺ and K⁺ ions.*

48. At point A, the membrane is permeable to Na⁺ ions, and thus the cytoplasmic side becomes transiently positive. Since, at B, the cytoplasmic side of the membrane is negative, electric current flows in a local circuit from B to A, and as a result the membrane at point B becomes permeable to Na⁺. At C, the resting potential has been re-established. The net result is a wave of positivity spreading along the inner surface. In unmyelinated nerve fibres and muscle cells this is the primary mode of action potential propagation.

2. *The cable effect.* An impulse is also conducted when an internal redistribution of charge generates the same wave of positivity on the internal surface of the plasma membrane even though a transfer of ions across the membrane does not occur. This mechanism of impulse propagation (known as the 'cable effect' — it has many of the features of a submarine

Figure 48. *The generation of a local circuit.*

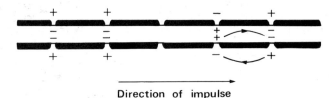

Direction of impulse

Figure 49. *The propagation of an impulse in a myelinated nerve fibre.*

cable) is improved if the capacity of the membrane to hold a charge can be reduced. This can be achieved by insulation, although to be sustained the impulse must be regenerated at regular intervals. In myelinated nerve fibres these conditions are provided for by an insulating myelin sheath that is interrupted at intervals by the node of Ranvier. The impulse in these fibres is thus propagated by a cable effect that is regenerated by a depolarizing influx of Na^+ at each node (Figure 49).

The myelin sheath surrounding the nerve fibre is composed of layer upon layer of plasma membrane (Figure 50) which belongs to the fibre's attendant Schwann cell. In keeping with its role as an insulator, this membrane is characterized by a high lipid content (about 80 per cent, by weight). As also may be expected of a membrane that is otherwise functionally inert, it has a low protein content and (as is seen on freeze-fracture) few intramembranous particles.

In many of the neurones that stimulate other neurones, and in all of those that stimulate striated muscle fibres, the chemical transmitter that causes depolarization of the target cell membrane is acetylcholine. However, although it is known that this transmitter binds to specific receptor sites on the target cell plasma membrane, the way in which this event provokes depolarization and causes the subsequent increase in the permeability of the membrane to Na^+ is unknown.

The plasma membrane and local anaesthetics

Analgesic. An agent producing pain relief.

The action of most local anaesthetics (and many analgesics) depends primarily upon their ability to become incorporated into the plasma membrane of sensory neurones and so prevent transmission of the nerve impulse. The potency of these substances is related to their lipid solubility, but it has not yet been clearly established if their effects concern only the membrane lipids. Knowledge of the way in which anaesthetics exert their effect at the molecular level is incomplete, but it is believed that they interfere and distort the organization of the plasma membrane in a general rather than a specific manner.

The transport of ions and small molecules across the cell

The maintenance of ionic balance across the plasma membrane is a continuous responsibility in all cells; it has been estimated that at any one

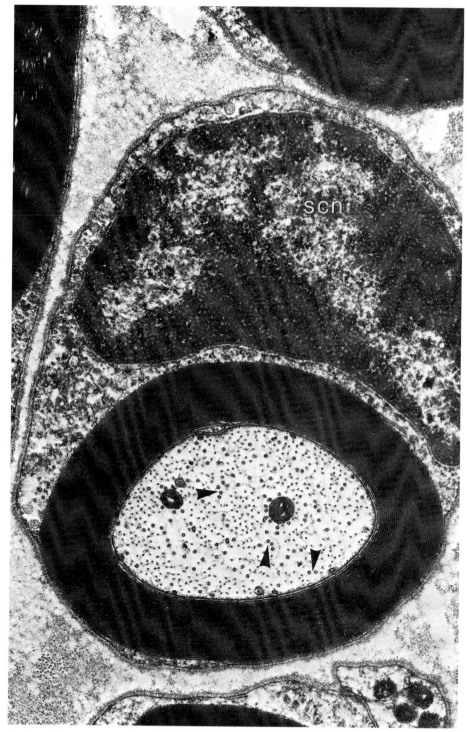

Figure 50. *An electron micrograph showing the axon of a myelinated nerve fibre in cross section. The plasma membrane of the Schwann cell forms about thirty concentric closely applied layers around the central axon. The innermost trilaminar membrane belongs to the neurone. Within the cytoplasm of the axon, mitochondria and cross-sectional profiles of both microtubules (arrows) and microfilaments are clearly shown. scn — Schwann cell nucleus. Magnification ×32 500.*

Courtesy of C. Peracchia, Department of Physiology, University of Rochester.

time this requires about 25 per cent of the cell's total available energy. In the cells of some tissues, however, the active transport systems of the plasma membrane have an additional transport responsibility, since they are also concerned with transporting water and ions across the cell from one extracellular compartment to another. This kind of transport often operates against steep concentration gradients, and, as a consequence, the total requirement for active transport is enormously increased. As a general rule, it seems that the efficiency of the individual pumping mechanisms in the plasma membranes concerned is not readily improved upon, and so the extra demand is usually accommodated by an increase in the number of sites at which active transport occurs.

Although it is now just possible to relate the microtransfer mechanisms themselves to microscopically identifiable sites within a membrane, the grosser modifications concerned with providing an increase in surface area are usually a major, and easily identified, feature in cells adapted for this kind of transport. A classic example is found in the columnar epithelial cells of the small and large intestines, where, in addition to water and ions, the small molecule products of luminal digestion (such as amino acids, simple sugars and fatty acids) are absorbed. As shown in Figures 51 and 52, the apical membranes of these cells are repeatedly infolded into finger-like microvilli to form a 'brush border'. In other tissues, such as the proximal convoluted tubule of the nephron and the

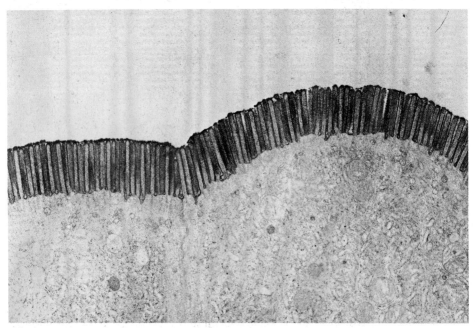

Figure 51. *The brush border of mucosal epithelium in the small intestine. This low-power electron micrograph shows the apical brush borders of two cells. Magnification ×2800.*

Figure 52. *The apical brush border of a mucosal epithelial cell at high magnification. Note that the plasma membrane of each microvillus has a distinct trilaminar appearance, and that on their external surfaces the outer laminae are finely fibrous. Interdigitating between each microvillus proper are the fibrous surfaces of adjacent microvilli, which are cut in grazing section. Magnification ×100 000.*

Courtesy of C. Peracchia, Department of Physiology, University of Rochester.

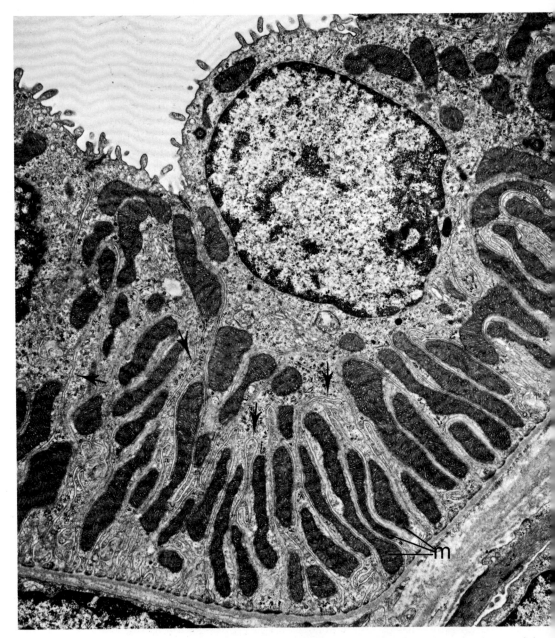

Figure 53. *A section across the wall of a distal convoluted tubule in the kidney. The plasma membranes of the lining epithelial cells are so extensively infolded (arrows) that many of the cytoplasmic compartments seen in this section belong to neighbouring cells. Aligned along these infolded membranes are large, well-developed mitochondria (m). The apical microvilli, it should be noted, are not particularly well developed.*

Courtesy of R. E. Bulger, Department of Anatomy, University of Massachusetts.

choroid plexus in the central nervous system, where the lining cells also have a major responsibility for the active absorption of water, ions and small molecules, there are similar elaborate microvillous borders.

In the distal convoluted tubule of the nephron the situation is rather different. Here the lining cells are required to transport Na^+ from the hypertonic fluid in the lumen out into the surrounding tissue fluid. They fulfil this requirement by actively pumping these ions across the basal boundary of the cell and allowing them to be replaced by those flowing in passively from the lumen (down the electrochemical gradient). As expected, the infolding of the plasma membrane in these cells is not particularly well developed on the apical surface, but it is very elaborate in the basal region. Indeed, as shown in Figure 53, in the basal region the lateral cell membranes are elaborately inflected and interdigitate extensively with those of the surrounding cells. Some indication of the local energy requirements of the active processes in these inflected membranes is given by the large numbers of well-developed mitochondria that are distributed in parallel alongside them.

Macrotransfer processes

The transport of macromolecules across the plasma membrane comes under the general heading of cytosis; endocytosis describes movement into the cell, and exocytosis movement into the extracellular space.

Endocytosis

Endocytosis includes phagocytosis, the process whereby macrophages engulf and remove cellular and non-cellular debris from the body (Figure 54). Phagocytosis is a widespread activity carried out under normal and pathological conditions by a system of cells called the mononuclear phagocyte system (distinguishing them from the multinucleate giant cells that arise in certain pathological conditions). This system includes the ubiquitous, freely mobile macrophages of the connective tissues and also the fixed phagocytic cells of the spleen, bone marrow, liver, and central nervous system. It also includes granular leucocytes.

Pinocytosis and micropinocytosis also come under the heading of endocytosis. The mechanisms involved in these processes are similar to those of phagocytosis but they are primarily concerned with the uptake of soluble materials dissolved in tissue fluids. They also have a rather different purpose, since they are usually concerned not so much with scavenging but with the absorption and transport of macromolecular materials directly of use to the body. Pinocytosis can be observed using phase contrast microscopy with living cells in culture, but micropinocytosis is, by definition, below the resolution of the light microscope. Micropinocytotic mechanisms are, nevertheless, capable of transporting large amounts of material and are, for example, primarily responsible

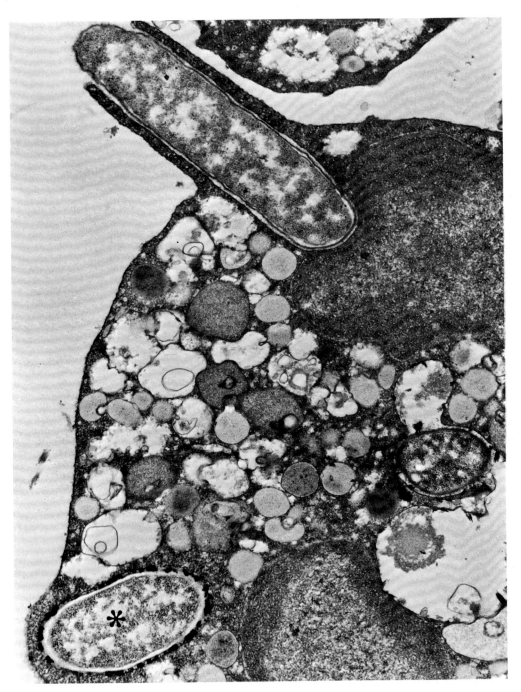

Figure 54. *The phagocytosis of a micro-organism by a polymorphonuclear leucocyte. At upper left, the cell boundary is extended and the bacterium is about to be enveloped.*

This preparation has also been treated histochemically to demonstrate the distribution of the lysosomal enzyme, acid phosphatase. At lower left, the phagocytosis of a second bacterium (∗) is complete; the presence of an electron-opaque precipitate around its periphery indicates that lysosomal enzymes (and, in particular, acid phosphatase) have gained access to the endocytotic vesicle.

Courtesy of D. F. Bainton, University of California.

for the reabsorption of proteins from the lumen and the nephron and for the uptake of thyroglobulin by the thyroid follicular cell (see page 183).

In outline, endocytosis involves the local invagination of the cell membrane and then the pinching off of the invagination to form a membrane-limited vesicle within the cell. In phagocytosis, the invaginating cell membrane is closely applied to the particle being taken up, so that the surrounding extracellular fluid is effectively excluded. In pinocytosis, on the other hand, soluble materials (and submicroscopic particles like ferritin) are often taken up along with the tissue fluid in which they are dissolved.

The terms 'fluid endocytosis' and 'absorptive endocytosis' are used to indicate the difference between substances that are taken up in free solution and those bound to the cell surface (Figure 55). In fluid endocytosis the concentration of the substance within the enclosed vesicle is the same as that in the extracellular environment. In absorptive endocytosis, on the other hand, the concentration of the substance being taken up depends upon its ability to bind to the cell surface. In some circumstances the amount of surrounding carrier fluid may be actively reduced by the invaginating vesicle collapsing just before it pinches off.

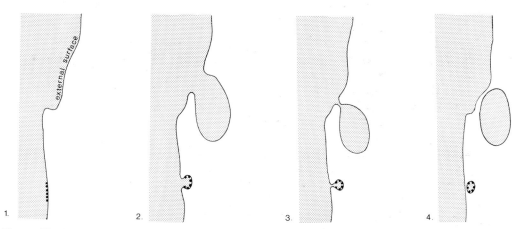

Figure 55. *Absorptive and fluid endocytosis. In absorptive endocytosis the substance inducing uptake binds to the cell surface and is selectively concentrated within the endocytotic vacuole. In fluid endocytosis the content of the vacuole is the same as that in the extracellular space.*

Absorptive endocytosis

Whether or not a particle is endocytosed by a cell depends primarily upon the nature of the particle surface. The surface features concerned may be generalized properties such as charge or hydrophobicity or they may be very specific molecular configurations for which there are specialized receptor sites on the cell surface. On fibroblasts, for example, there are receptors specifically concerned with the uptake of cholesterol-containing low-density lipoproteins. The uptake of these complexes plays an important role in the regulation of cholesterol metabolism that is well illustrated by the condition of hypercholesterolaemia. In this condition there is often a gene-linked deficiency in the number of these surface receptors, and as a result there is a marked reduction in lipoprotein uptake (see page 184 for further discussion).

Phagocytosis and the immune response

Although mononuclear phagocytes represent a first line of defence against invading agents like bacteria, their ability to recognize these agents directly is limited. The responsibility for recognition thus often falls to the more discriminating cellular components of the immune system. These components will identify certain molecular features of the bacterium surface as 'foreign' (i.e. antigenic — see page 41) and they are then able to generate a complex response that leads to the production of specific antibodies (and their associated serum factors known as 'complement') that will identify and bind to these antigenic features. Bacteria coated with bound antibody and complement are said to be 'opsonized' and as such they are readily identified and engulfed by mononuclear phagocytes. This is because the phagocytes bear specific receptors on their surface that are able to identify these bound molecules. The usefulness of this complex system is illustrated by strains of bacteria in which there are non-capsulated, non-pathogenic forms, which are readily identified and phagocytosed, and capsulated forms, which are more pathogenic because their capsules serve to prevent them being recognized by the phagocyte surface. Subsequently, when antibodies are produced against their capsule surface, these pathogenic forms also become identifiable and they, too, are phagocytosed.

In regard to the endocytotic mechanism it should be noted that, when membrane fusion occurs, the continuity and integrity of the cell boundary is always maintained. It is also worth emphasizing that the material taken up by this process does not gain access to the cell cytoplasm per se, but remains sequestered within a limiting membrane derived from the plasma membrane. The subsequent fate of this kind of endocytotic vesicle is indicated in Figure 55 and will be described in detail in Chapter 6.

Exocytosis

In outline exocytotic events are essentially the same as those of endocytosis in the reverse order. They are the mechanism whereby most exocrine and many endocrine secretory cells release their secretory products, cholinergic nerve fibres release their stores of acetylcholine, and osteoclasts (which can degrade bone matrix extracellularly) secrete lysosomal enzymes. All of these processes require the transported material to be

Complement

Opsonize. From the Greek 'opsonein', 'to prepare food'.

Pathogenic. Causing disease.

Exocrine

Endocrine

Cholinergic nerve

Osteoclasts

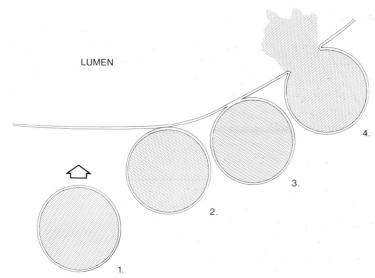

LUMEN

Figure 56. *The exocytosis of secretory product. When stimulated, the secretion granule moves to the cell periphery (1, 2), the granule and plasma membranes fuse (3), and the content is released (4).*

packaged in the cytoplasm within a membrane-limited vesicle, such as a secretory granule or a lysosome. A demand for the release of content (e.g. a stimulation to secrete) induces the vesicle to move to a prescribed region of the cell boundary (in exocrine cells it is usually the apical membrane) where the vesicle membrane fuses with the plasma membrane (Figure 56). At the present time our understanding of the molecular events which direct the vesicle and bring about membrane fusion is fragmentary. However, as with endocytosis, it is clear that the transfer of material is accomplished while the cell boundary remains intact. A consequence of exocytosis is the addition of intracellular membrane to the plasma membrane; when this occurs repeatedly and becomes a significant contribution, it seems to be followed by a compensatory endocytotic withdrawal of membrane from the surface and back into the cell.

Apical membrane

The transport of macromolecules across the cell

In some situations, macromolecules are transported across the cell (i.e. from one extracellular compartment to another), and in these circumstances the cell rather than the plasma membrane represents the selective barrier. Good examples of this kind of transport are found in the blood capillaries (Figure 57).

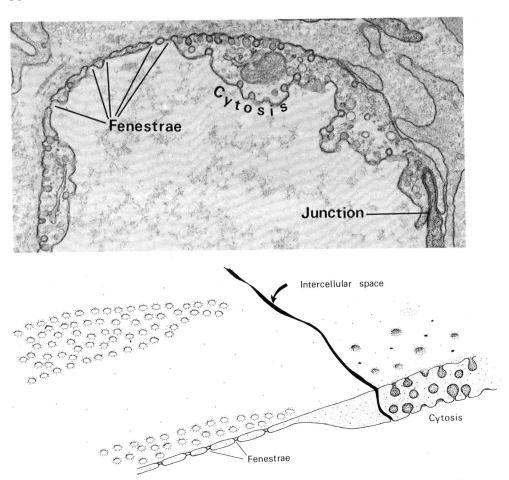

Figure 57. *The different routes that can be taken by macromolecules across the capillary wall.*

Electron microscope studies employing thin sections, freeze-fractured replicas, and electron-opaque tracers of different size* suggest that macromolecular substances may cross the capillary wall by at least three different routes. In the capillaries of some tissues, transfer between the endothelial cells that line capillary walls may represent a major pathway, although in other tissues (and especially in those of the central nervous system, where a 'blood–brain barrier' exists) these intercellular spaces are bridged by impermeable junctional elements (see below). In the capillaries of most tissues, in addition to any transport that may occur

---

*The available tracers are either themselves electron-opaque or their presence is demonstrable electron-histochemically. Ferritin (mol. wt. 500 000, molecular diameter 11.0 nm) and haemoglobin (mol. wt. 64 500, md 6.4×5.0×5.0 nm) were amongst the first tracers to be used, but the introduction of catalase (mol. wt. 24 000), horseradish peroxidase (mol. wt. 40 000, md 5.0 nm) and its much smaller derivative, microperoxidase (mol. wt. 19 000, md 2.0 nm), has now extended the range of probes available. Although the colloid tracer, lanthanum hydroxide, is of unknown molecular size, it is now being used widely, for in some situations it alone is small enough to penetrate the attenuated extracellular space.

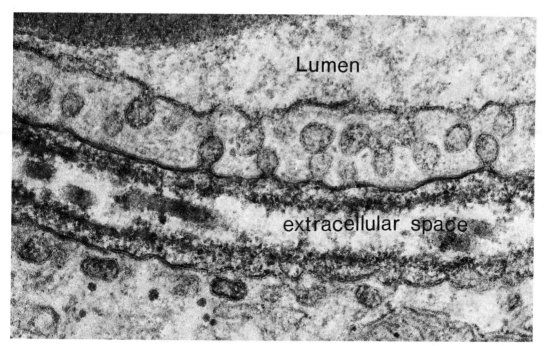

Figure 58. *A high-magnification view of an unfenestrated capillary, showing abundant evidence of cytosis. Courtesy of N. and M. Simionescu, Section for Cell Biology, Yale University.*

between the endothelial cells, transport also occurs across the cell boundary proper, either by means of cytosis or via window-like pores ('fenestrae').

Transport by cytosis depends upon the formation, on one side of the cell, of small (60 to 70 nm diameter) endocytotic vesicles, and their transfer to the other side, where they release their content by exocytosis (Figures 58 and 59). Of the routes taken by macromolecules across the capillary wall, transport via individual cytotic vesicles is probably the slowest, although it is likely that, as a result of several vesicles fusing, transient through-channels across the cell may also be produced. Fenestrae (Figure 60), on the other hand, are permanent rather than transient features of the capillary wall, and they represent patent (60 nm diameter) pores in the endothelial cell. Each pore is bordered by plasma membrane, and is usually bridged by a thin diaphragm; the diaphragm is presumably composed of peripheral plasma membrane components that extend across the pore orifice.

Comparative studies on the number and distribution of cytotic invaginations and fenestrae in the capillaries of different tissues show that, although all three alternative routes across the capillary wall are probably available in many situations, in others either cytotic invaginations or fenestrae predominate. In these situations, these features usually comprise a fairly consistent and even characteristic proportion of the total

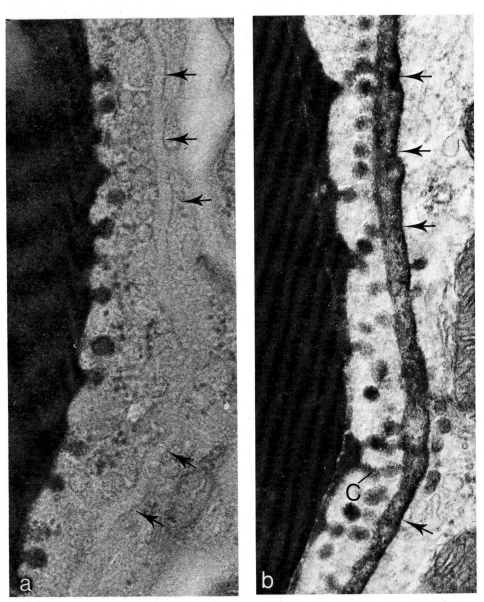

Figure 59. *The transport of a tracer molecule (microperoxidase – identified histochemically) across the capillary wall by cytosis. In (a), the tracer has moved from the lumen into the cytotic invaginations on the luminal surface, while in (b) (within 60 seconds) it has been transported across the cell and expelled into the surrounding extracellular space. C indicates a point at which the fused vesicles provide a through channel. In both micrographs the position of the extracellular space is indicated by arrows. Magnification ×42 000.*
    Courtesy of N. Simionescu, M. Simionescu and G. E. Palade, Section for Cell Biology, Yale University.

available surface area. In heart muscle capillaries, for example, where fenestrae are virtually absent, there are about 90 cytotic invaginations per μm^2, while in those of diaphragm muscle there are about 80. In the capillaries of the small intestine, on the other hand, where fenestrae may comprise as much as 30 per cent of the total surface area, the

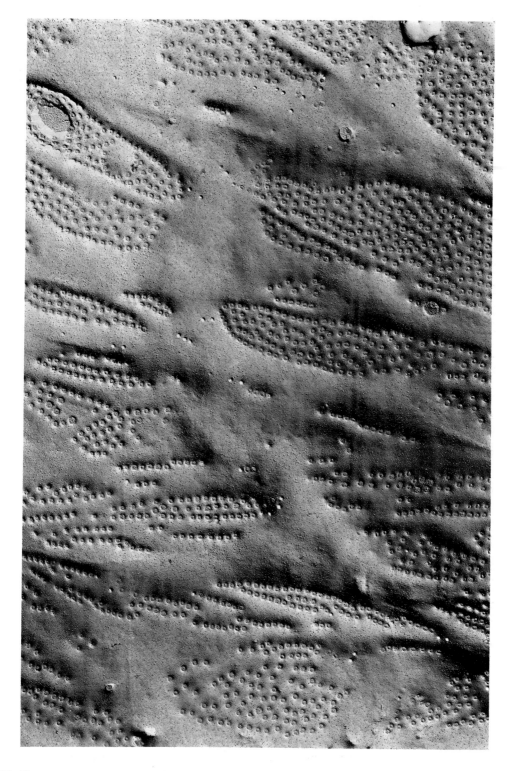

Figure 60. *The distribution of fenestrae, shown in a replica of a freeze-fractured capillary wall. Magnification ×15 750.*
  Courtesy of C. Peracchia, Department of Physiology, University of Rochester.

number of cytotic invaginations is reduced to about $10/\mu m^2$. In the pancreas, where the capillaries have rather fewer fenestrae than in the intestine, the number of cytotic invaginations is correspondingly greater (about $20/\mu m^2$). Unfortunately, although this differential distribution clearly implies that in any particular tissue the different transport routes make rather individual contributions to the overall transport requirement, the extent to which they represent alternative or even selective pathways is not known.

THE PLASMA MEMBRANE AND CELL ATTACHMENT

The cells of lining epithelia and solid tissues are held together by a combination of the generalized property of cell surface adhesiveness and a variety of specialized areas which comprise well-defined junctional elements.

The molecular basis of cell surface adhesion is unclear, although it is almost certainly a property of the exposed carbohydrates of the plasma membrane glycoproteins. The attachment can be very specific, for, although some kinds of cell will adhere to one another at random, and will even stick with tenacity to non-biological substances such as glass, others, in a suspension of mixed cell types, will 'sort out' selectively and adhere only to their own kind.

Cell junctions

Labelling experiments indicate that over most of the free cell surface the plasma membrane proteins are more nearly analogous to freely floating icebergs than packed ice-floes. However, in the junctional areas of the cells of solid tissues, the latter analogy is more appropriate, for in these regions it appears that the protein components of the cell membrane are indeed arranged in closely packed aggregates. In addition, they may be anchored in position either by filaments arising from within the cell, or by their connection with similar components on neighbouring cell surfaces. Junctional elements therefore represent discrete boundaries, which, in addition to their roles in attachment and controlling intercellular permeability (see below), divide the free cell surface into separate domains, Within each domain many membrane proteins probably have free lateral mobility, and the presence of a junctional element (especially when zonular in form) thus allows different regions of the cell surface to be structurally separate and functionally specialized. In polarized secretory cells, like those of the exocrine pancreas, for example, the receptors concerned with receiving hormonal signals from the bloodstream are almost certainly restricted to the basal and lateral membranes, while those plasma membrane components that participate in exocytosis (i.e. the release of secretion) are probably confined to the apical surface. The two domains are separated by extensive junctional elements near the luminal border (Figure 61).

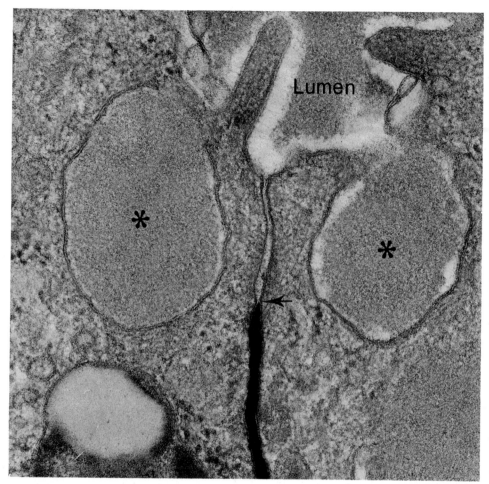

Figure 61. *An electron micrograph of the apical borders of two adjacent cells in the exocrine pancreas. These cells release their protein secretion by exocytosis into the gland lumen. Two secretory granules at the point of releasing their contained secretion are shown (∗).*

Between the two cells is a tight junction (arrow). Its impermeability is demonstrated by the electron-opaque tracer, lanthanum, which has penetrated through the intercellular space from the bloodstream. Magnification ×80 000.

Courtesy of D. S. Friend and N. B. Gilula, University of California and Rockefeller University.

The classification of junctional elements

Several different approaches are required to evaluate the nature of a junctional element. These include its examination in thin section, freeze-cleaved replicas and negatively stained preparations, the use of electron-opaque tracers of varying size (see footnote on page 90), and the measurement of the electrical resistance between the participating cells using intracellular electrodes.   *Negative staining*

Three major kinds of cell junction are recognized. They are the tight junction, the desmosome, and the gap junction (or nexus). In describing them, reference is usually made to their form and extent. Thus, if there is no intercellular space at the junction, the adjective 'occludens' applies; alternatively, if membranes closely approximate but do not close the intercellular space, 'adhaerens' applies. A 'macula' indicates a spot or small localized area; a 'fascia', a sheet-like area; and a 'zonula', a belt-like zone.

The tight junction

This kind of junction occurs as a continuous belt of intimate contact (zonula occludens) between epithelial cells, and serves to reduce or prevent intercellular transport through the extracellular space between adjacent cells. The permeability of tight junctions is different in different tissues, but as a rule it prevents the passage of all experimental tracers (Figure 61) and in many tissues it precludes even the passage of water and ions. In thin sections, the plasma membranes of tight junctional areas appear as a pentalamellar structure, with the two outer leaflets of the participating membranes completely obliterating the intercellular space. Freeze-cleaved replicas expose the tight junction as a continuous lacework of grooves and ridges running without a break to completely encircle the cell. In epithelia in which the junction is particularly impermeable, the lacework is complex (Figure 62); in other tissues, such as those of the proximal convoluted tubule of the kidney, where the junctions are more permeable, the lacework consists of only one or two strands.

The desmosome

Desmosomes are restricted, button-like areas (macula adhaerens) which are concerned with cell attachment but are not involved in either intercellular or intracellular transport. In thin sections of the desmosome (Figure 63), the parallel plasma membranes of adjacent cells are separated by a 25 to 30 nm space that often contains an electron-opaque, carbohydrate-rich material. On the cytoplasmic surfaces of the membranes, fibrous plaques act as attachment sites for radiating, 10 nm diameter fibrils called 'tonofilaments'. The tonofilaments probably serve to dissipate the stresses exerted at the macula throughout the cytoplasm. The protein, desmin, is a major component of these fibres, although there is some evidence to suggest that they may become closely associated with the contractile protein, actin. Desmosomes are well developed in

Epidermis

epithelia where mechanical stresses occur, such as the skin epidermis; in certain skin diseases where there is a premature shedding of dead epidermal cells (squames), reduced desmosomal development may be the cause.

Intercalated discs

Zonula and fascia adhaerens are found in the intercalated discs of cardiac muscle fibres. They differ from desmosomes in that the intercellu-

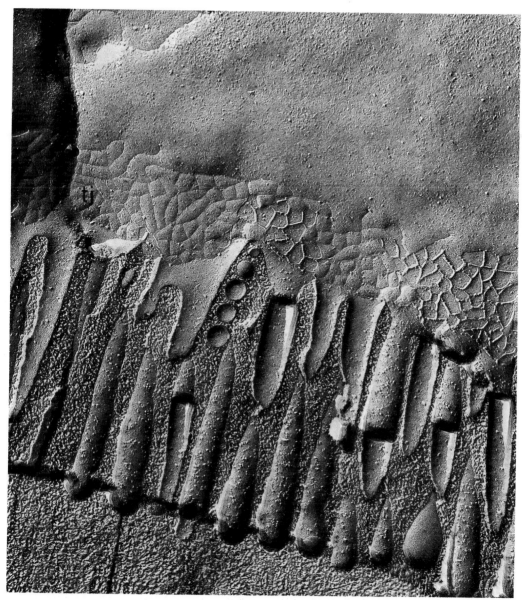

Figure 62. *A replica of freeze-fractured mucosal epithelium from the duodenum. Near the apical margin and parallel to the microvillous border there is an extensive tight junction (tj). Its zonular (belt-like) distribution and the complexity of the interconnecting furrows and ridges are clearly shown. Magnification ×73 000.*
Courtesy of D. S. Friend and N. B. Gilula, University of California and Rockefeller University.

lar space is smaller (15 to 25 nm) and it contains little or no condensation of electron-opaque material. In this situation it is probable that the actin filaments insert into the plasma membranes via condensations containing the filamentous proteins desmin and alpha actinin (see page 232).

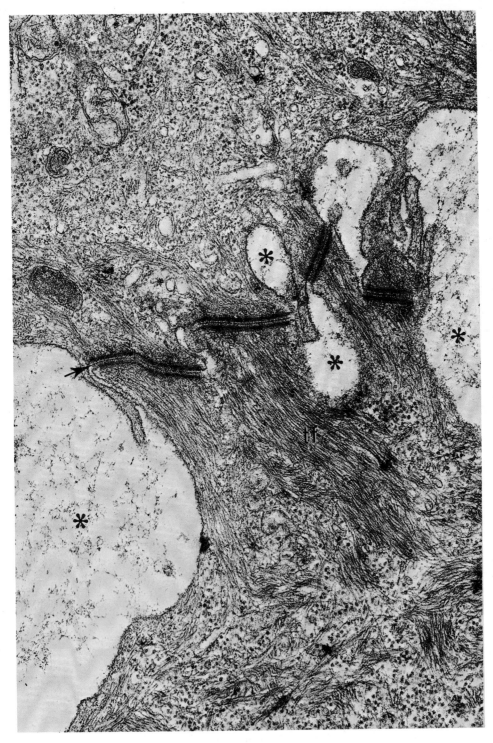

Figure 63. *Desmosomal attachments between two cells in the skin epidermis. Electron-opaque condensations on the cytoplasmic side of the membrane at the points of attachment of the tonofilaments (tf) are well shown. The form and arrangement of the filaments themselves are best seen in the lower cell. Intercellular material is indicated by the fine linear condensate between the cells (arrow); the extracellular spaces are indicated by asterisks. Magnification ×18 000.*

Gap junctions

This kind of intercellular junction is primarily concerned with cell-to-cell communication. In thin sections, it appears as a macular or even fascia adhaerens in which the participating plasma membranes closely approach one another but do not fuse. Typically, the intercellular space between the membranes is about 2 to 3 nm, and although it can be penetrated by some intercellular tracers, such as lanthanum, it is impermeable to others, such as horseradish peroxidase. Where in thin sections the section plane grazes the cell surface and shows a gap junction en face, the plasma membranes in the junctional area are seen to be composed of 7 to 8 nm diameter particles. These components are often in an hexagonal array, and high-resolution and X-ray diffraction studies suggest that they are in fact composed of six subunits arranged to form a cylinder about a central hydrophilic channel (Figure 64). It is believed that on the opposing faces of the participating plasma membranes these components are in register so that their hydrophilic channels are confluent.

Freeze-fractured replicas provide a good indication of the surface distribution of gap junctions, and show that they may vary from extensive, belt-like zones, as in liver cells (Figure 65), through discoid maculae, as in ovarian granulosa cells (Figure 66), to the single strands of particles seen between some kinds of cells in culture. For any particular cell type or location, however, the form of the gap junction is usually fairly characteristic. Although on freeze-fracture the particulate components of the

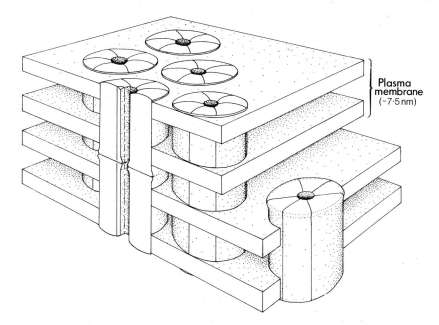

Plasma membrane (~7·5 nm)

Figure 64. *A reconstruction of the organization of a gap junction, compiled from observations made on thin sections and freeze-fractured replicas. After D. Goodenough.*

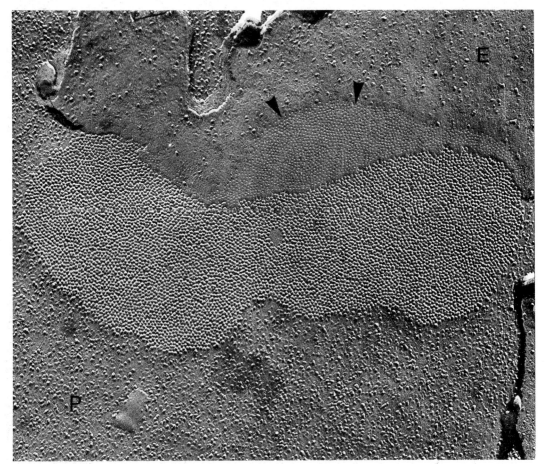

Figure 65. *A replica of a freeze-fractured gap junction from the liver, showing its elongated plaque-like form. On the P face in the junctional area, the intramembranous particles are distributed in a tightly-packed, polygonal array. This distribution is mirrored by the distribution of the pits (which are indicated by arrows) on the E face, the other leaflet of this same membrane. Magnification ×96 000.*
Courtesy of N. B. Gilula, Rockefeller University.

gap junction appear rather smaller (4 to 6 nm) than those seen in thin sections, they have the same distribution, and there is often a clear indication of a discrete central (1.5 to 2.0 nm diameter) pore.

Our understanding of the functions of gap junctions is very incomplete. There is no doubt, however, that they are widely distributed throughout most tissues, and in all of the instances examined so far they have been shown to be highly permeable. Their special characteristic is that they allow a ready exchange of ions between the cytoplasmic

Figure 66. *(Opposite.) Gap junctions between ovarian granulosa cells, shown (a) in thin section and (b) in a freeze-fractured replica. The tissue prepared for thin section microscopy had been previously immersed in lanthanum, which has outlined (arrows) the closely opposed plasma membranes in the gap junctional area. The freeze-fractured preparation is photographed at a higher magnification, and shows both the macular form of this junction and the orderly array of its intramembranous components. Magnification: (a) ×40 000; (b) ×139 000.*
Courtesy of D. F. Albertini and E. Anderson, Harvard Medical School.

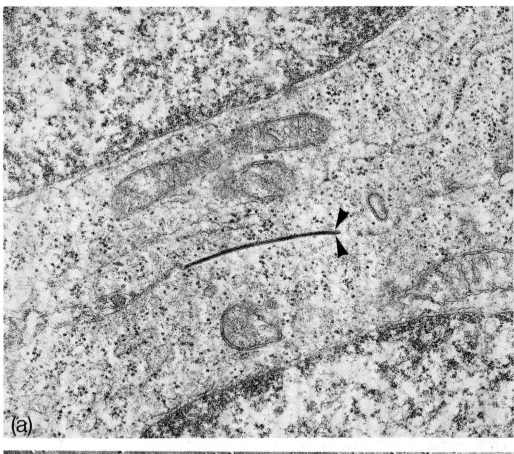

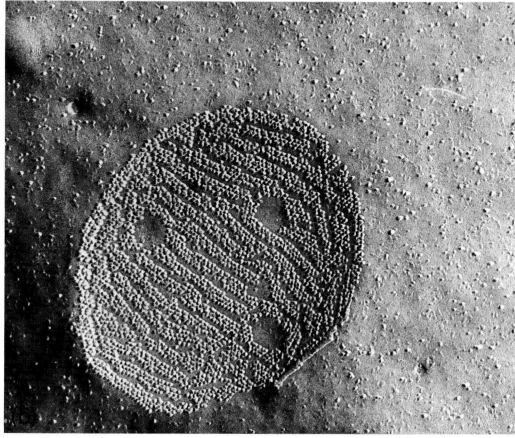

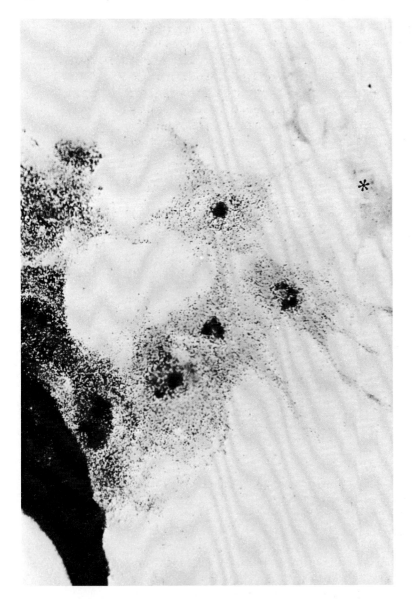

Figure 67. *Communication between cells in culture shown by the transfer of labelled uridine nucleotides. Donor cells previously incubated with radioactive (tritiated, ³H) uridine were co-cultured with unlabelled cells for three hours before the preparation was fixed and prepared for autoradiography. Under these conditions the donor cells contain radioactive uridine, which has been incorporated into uridine nucleotides and RNA; when these cells establish junctional contact with the cells of the recipient preparation, the nucleotides (but not RNA, as it is too large) can be transferred. The nucleotides may then remain in the cytoplasm of the recipient, or they may be incorporated into its RNA (and label its nucleolus), or they may be handed on, via further junctional contacts, to neighbouring recipient cells.*

In the autoradiograph shown here, the donor cell (heavily labelled, lower left) has transferred labelled nucleotides to a chain of recipients. Although all of the recipients have incorporated sufficient nucleotide for their nucleoli to become labelled, it is clear that, as they become further removed from the donor, the amount of label they each contain diminishes. At upper right (*) is a cell without contact with the donor and without label, indicating that the radioactive nucleotide is available only to cells that are coupled to a labelled donor. Almost certainly, this transfer occurs via gap junctions. Magnification ×1125.

Courtesy of J. D. Pitts, Department of Biochemistry, University of Glasgow.

compartments of the cells they join. Between cells in culture (Figure 67), gap junctions are probably able to transfer molecules as large as nucleotides, although it is not known whether this is a property of gap junctions in general. Also in cultured cells it has been shown that gap junctions can form rapidly (within seconds) without requiring protein synthesis; this suggests that, when required, gap junctional plaques may arise simply by the subunits resident in the participating plasma membranes aggregating in apposition to one another. Their formation under these conditions appears to be quite specific, since it only occurs between certain kinds of cell.

Nucleotides

Together, these properties suggest that gap junctions play an important role in intercellular communication. In many tissues (and especially in smooth muscle, where the term 'nexus' has often been applied to them) it is probable that their permeability to ions allows them to propagate action potentials; elsewhere, it is conceivable that other kinds of signal molecule, such as cyclic nucleotides, may be transferred. The positioning of gap junctions between differentiating cells in the developing embryo, and between the constituent cells of tissues, like the muscle and exocrine gland cells, in the adult, suggests that gap junctions may be important in situations where groups of cells are required to act in concert. However, although this suggestion is attractive, not least because it has also important implications for the role of gap junctions in the regulation of cell proliferation, it remains to be confirmed experimentally.

Junctional complexes

Between adjacent cells in transporting and secretory epithelia there is commonly a series of junctional elements (Figure 68). These junctional complexes usually include a tight junction near the luminal border and desmosomal elements situated more basally.

Frequently, either gap junctions or fascia adhaerens (desmosomes with a sheet-like distribution) occur in an intermediary position. Together, these elements provide all of the intercellular relationships necessary for a tissue required to provide a strong sealed lining.

The attachment of cells to the substratum – the basal lamina

Although glycoprotein components predominate on the external surface of all cells, in some areas their development becomes very extensive and it is clear that a large proportion of them are quite separate from the integral membrane structure. This is especially evident where epithelia are associated with a basal lamina, a substratum that provides the interface between them and the underlying connective tissue. A basal lamina usually contains connective tissue elements, such as reticular and

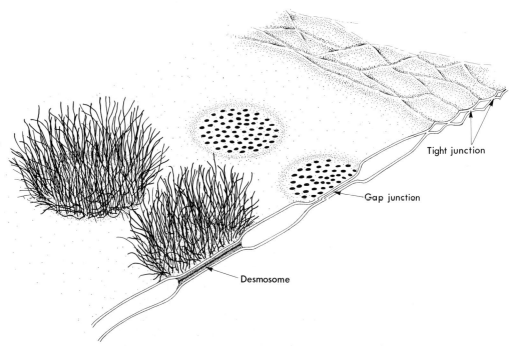

Figure 68. *A reconstruction of the kind of junctional complex that occurs between adjacent cells in a lining or glandular epithelium. At upper right, near the luminal border, there is a tight junction (compare with Figure 62), towards lower left there is a desmosome (compare with Figure 63). Between them are plaque-like gap junctions. For simplicity, tonofilaments are shown only in the upper cell.*

Acinus

Plaque. A patch or spot on a surface.

collagen fibres, but the bulk of it is composed of glycoprotein that is derived from the epithelial cells.

The basal lamina of most epithelial layers is important for support and attachment, and during development it is also believed to play a role in their organization. For example, in the development of exocrine glands it is required for the formation of the lobular arrangement of the secretory acini. Its role in attachment is indicated directly by the presence of hemidesmosomes – plaque-like attachments on the basal membrane of the epithelial cells that, together with their associated cytoplasmic tonofilaments, are in all respects the morphological counterpart of desmosomes, which occur between adjacent cells. In general, the basal lamina is thought to be freely permeable to even large molecular species, although in the capillary wall (see Figure 58) and in the renal glomerulus it is believed to play some part in selective filtration.

5

THE NUCLEUS

GENERAL FEATURES

Interphase nuclei (i.e. nuclei that are not dividing) are surrounded by a nuclear envelope and contain a variously condensed chromatin network together with one or more nucleoli (Figure 69). Their role as a store of genetic information and a controlling centre concerned with the selective expression of the stored information is well established, but many important molecular details remain to be defined. For this reason it will be easiest for us to discuss first the nucleus in relation to its major functions. Then we will turn to the poorly understood significance of its structural appearance.

FUNCTIONS OF THE NUCLEUS

To obtain a proper understanding of nuclear function, it is necessary to provide a satisfactory account of the following:

1. The molecular basis for the storage of hereditary information.
2. The mechanism whereby hereditary information is duplicated at cell division.
3. The way in which genetic information in the nucleus is expressed and able to influence events in the cytoplasm.
4. The controlling mechanisms that dictate which portion of the total genetic information of the organism (the 'genome') is expressed in any particular cell at any particular time. *Genome*
5. The ways in which the controlling mechanisms themselves are influenced by changes in the intra- and extra-cellular environment.

THE PROKARYOTIC SYSTEM

In molecular terms, our understanding of how genetic information is stored, duplicated and expressed (items 1 to 3 above) depends almost

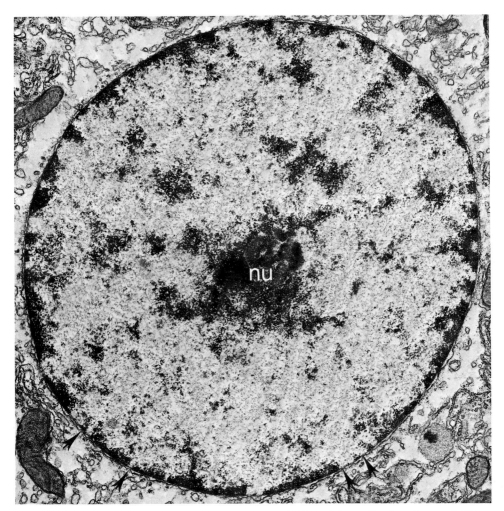

Figure 69. *The interphase nucleus of a liver parenchyma cell. Most of the nucleoplasm is occupied by an expanded chromatin network; condensed chromatin is confined mainly to the periphery, below the nuclear envelope. In the central region lies a single nucleolus (nu). The arrows indicate some of the many nuclear pores. Magnification × 10 000.*

Prokaryote

Eukaryote

Genetic code

Polynucleotide chain

entirely upon studies using bacteria (prokaryotes). These studies provide a useful working model against which the more complex and much less easily studied eukaryotic system can be compared.

In bacteria, genetic information is embodied within the triplet sequence of nucleotide bases (the 'genetic code') that run along the double polynucleotide chains of deoxyribonucleic acid (DNA) molecules. The two polynucleotide chains have a complementary sequence of bases and are arranged in a double helix about a common axis. When a bacterium divides, precise copies of the parental DNA are produced, because each of its polynucleotide chains is used as a template for the

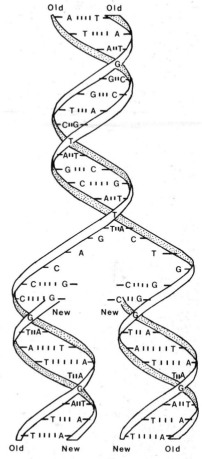

Figure 70. *The replication of DNA. A, T, G and C are the nucleotide bases adenine, thymine, guanine and cytosine, respectively. After C. A. Villee.*

synthesis of a new complementary chain (Figure 70). Two identical double helices, each containing one parental and one 'new' chain, are thus provided for the progeny.

Most studies on the expression of genetic information in prokaryotes have concentrated upon the control of protein synthesis, when the genetic code dictates the ordered sequence of the amino acids in a polypeptide chain. The group of nucleotide sequences responsible for the synthesis of any single polypeptide is said to constitute a 'structural gene'. The so-called 'central dogma' of molecular biology maintains that genetic information is always transferred *indirectly* from the structural gene to the protein-synthesizing machinery. The intermediary is identified as 'messenger ribonucleic acid' (mRNA). The sequence of nucleotide bases in mRNA mirrors precisely the nucleotide sequence of the structural gene, because when the mRNA is synthesized (using the *Polymerase* enzyme, RNA polymerase) DNA is again used as a template. This part of the process of information transfer is referred to as 'transcription'. *Template*

The events which employ the information encoded in the nucleotide sequence of an mRNA for the synthesis of a polypeptide chain comprise the process of 'translation'. They require the participation of ribosomes—

discrete, ribonucleoprotein particles that provide the molecular framework upon which each triplet sequence in the mRNA is able to dictate
the orderly sequential attachment of the amino acid residues.

Escherichia coli   In bacteria the two processes of information transfer (transcription
and translation) are tightly coupled. Indeed, as shown in Figure 71, the
information encoded within an mRNA molecule may be translated by its
associated ribosomes even while its transcription on the DNA is being
completed. Within minutes of being made, the mRNA is usually
degraded.

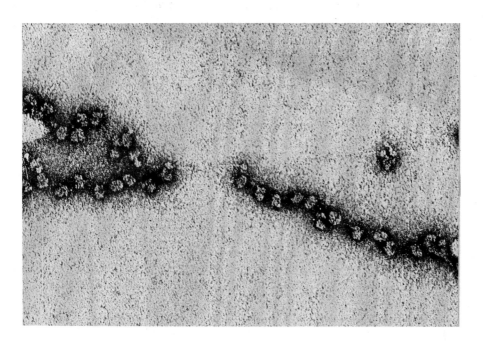

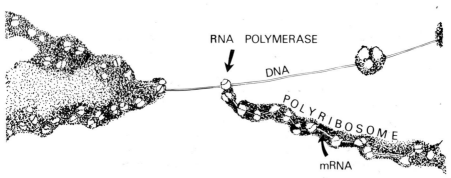

Figure 71. *Transcription and translation in the bacterium* Escherichia coli. *The RNA polymerase molecule (7.5
nm diameter) is seen transcribing a portion of the DNA; the mRNA produced is being translated by its associated
ribosomes. To obtain this preparation, actively growing bacteria were lysed (burst) in water and their extruded
contents spread onto an electron microscope grid. The preparation was then treated with a heavy metal stain
that shows the separated components in negative relief. Magnification ×160 000.*
  Courtesy of B. A. Hamkalo, Department of Molecular Biology and Biochemistry, University of California.

THE EUKARYOTIC SYSTEM

Mammalian cells typify the eukaryotic type of organization; in them, certain fundamental features of the prokaryotic system are clearly in evidence. Thus genetic information is carried by the same triplet nucleotide code on DNA, and mRNA is the intermediary produced by transcription to direct protein synthesis. There are, however, also major differences between the two systems. Structurally, the most dramatic are due to the presence in the eukaryote of the nuclear envelope, since this boundary serves to sequester transcription within the nucleus and confine translation to the cytoplasm.

Eukaryotic chromosome structure

Eukaryotic chromosomes consist of chromatin, a complex which, in addition to linear DNA molecules, contains nuclear proteins and small amounts of RNA. The nuclear proteins of chromatin include histones (small proteins of molecular weight less than 20 000, typically rich in the basic amino acids, lysine and arginine) and so-called 'non-histone', acidic proteins.

In the electron microscope, in conventional thin sections, chromatin appears to be made of myriads of fine, unbranched fibres without any preferred orientation (see Figure 88). These profiles have been difficult to analyse experimentally, and for meaningful studies of chromosome substructure it has been necessary to develop other methods of preparation. The most successful of these methods involves extracting condensed chromosomes from dividing cells (or chromatin-rich fractions from interphase nuclei) and then drying them down, whole, for examination in the microscope. Using this kind of preparation together with highly specific 'nucleases' (enzymes able to cut the DNA of the chromatin into short lengths), it has now been established that interphase chromatin (at least in heterochromatic regions – see page 124) consists of beaded strands, in which the linear double helix of DNA exists as a 2 nm fibre both encircling and connecting each bead (Figure 72). The beads are called 'nucleosomes', and each of them is associated with about 200 pairs of nucleotide bases along the DNA strand. The main body of each nucleosome is made up of a tightly ordered group of eight histone molecules.

In the condensed chromosomes of mitosis and meiosis, the chromatin appears to be composed of a mixture of closely infolded 10 and 30 nm diameter fibres (Figure 73). These fibres presumably consist of the nucleosome-associated 2 nm DNA fibre supercoiled to varying degrees of complexity.

DNA Helix ----

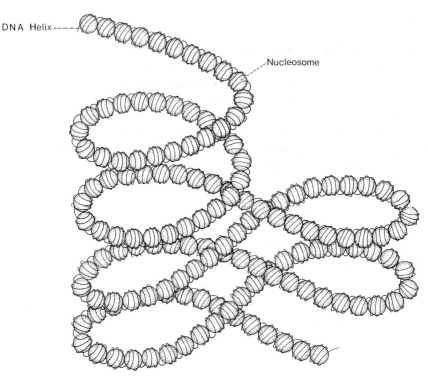

Nucleosome

Figure 72. *A model of the probable arrangement of DNA and its associated histones within the nucleosome subunits of the eukaryotic chromosome. Each group of histones forms an oyster-shaped complex, and has a 200 nucleotide length of the DNA strand wound about it.*

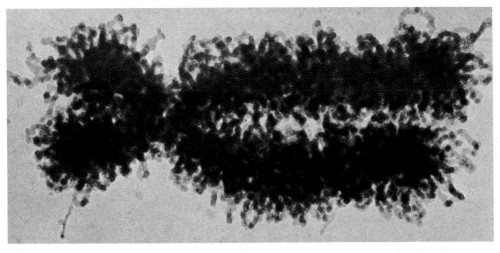

Figure 73. *An electron micrograph of a whole mounted human chromosome 12 showing the two chromatids composed of fibrils 30 nm thick.*
Courtesy of E. J. DuPraw, School of Medicine, University of Maryland.

Eukaryotic DNA replication

The structure of the eukaryotic chromosome is considerably more complex than that of prokaryotes, but in its basic essentials the process of DNA replication appears to be the same in the two systems. Thus, at the beginning of DNA synthesis, in both prokaryotes and eukaryotes, the DNA helix unwinds and the two complementary sequences of nucleotide bases become available for DNA polymerase enzymes to employ each of them as templates. In prokaryotes, unwinding begins at only one site on the DNA chain – the 'initiation site'. In the eukaryote, however, the unwinding process begins at multiple initiation sites along the chromosome length. Once unwinding begins, it extends in both directions away from each initiation site, and, as it proceeds, the synthesis of new, complementary strands occurs. The initiation of replication at multiple sites in the eukaryote has the advantage of reducing the time required for DNA replication. Otherwise, because DNA polymerases proceed rather slowly, the time taken to replicate an entire chromosome would be very much greater.* In early development, when cells divide more rapidly than in the adult, the number of sites at which unwinding is initiated is significantly increased.

THE REGULATION OF NUCLEAR ACTIVITY

In eukaryotes, the mechanisms regulating nuclear activity (items 4 and 5 on page 105) are complex and in all respects poorly understood. To simplify our account of them, they can be conveniently separated into those concerned with controlling DNA replication (and cell division) and those concerned with controlling gene expression in the interphase nucleus.

The control of cell division

During the development of the embryo, and in certain adult tissues, large numbers of cells divide. In normal tissues the number of divisions is closely regulated, but in malignant cancer cells proliferation is uncontrolled. In mitotically active adult tissues, such as the germinative *Epidermis* layers of the skin epidermis or the 'stem cells' of bone marrow, the intervals between each mitosis may be as short as eight hours. In other *Stem cells* tissues, however, they may be very much longer; in nervous tissue, for

---

* The DNA strand of human chromosome 13 is estimated to be about 30 mm long and there is a nucleotide base pair every 0.3 nm. DNA polymerases have been estimated to proceed at about 40 bases per second.

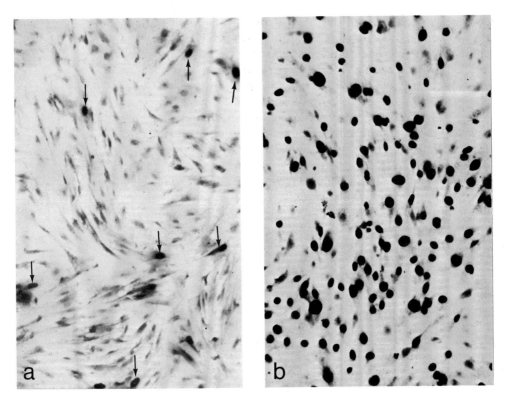

Figure 74. *Views at low magnification of cells growing in culture, showing the effect of stimulating them to divide by increasing the concentration of serum in the culture medium. In (a), the cells were acclimatized to low serum conditions. They were then incubated with radioactive thymidine for 12 hours and prepared for autoradiography. During the 12-hour interval only a small number of nuclei (arrowed) incorporated label, i.e. entered S phase.*

In (b), cells acclimatized to a low serum concentration were stimulated by raising the serum level from 0.1 to 5 per cent. Eighteen hours later, still in this serum, they, too, were incubated with radioactive thymidine for 12 hours. Their autoradiograph shows that during this time more than 80 per cent of them had begun to divide. Both magnifications ×200.

example, cells present in the newborn may remain without dividing for the lifetime of the individual.

Some cells thus divide repeatedly; others, once they have divided, enter a prolonged period of interphase, and in so doing normally 'differentiate', taking on the features characteristic of their tissue type (Figure 78). Although in adult tissues most divisions occur in undifferentiated stem cells, many differentiated cells (such as *Lymphocyte* lymphocytes and liver parenchyma cells), when appropriately stimulated, are also capable of further mitosis. The mechanisms and signals involved in the control of cell division in these different circumstances are an important aspect of nuclear regulation, and they are clearly of special interest to our understanding of carcinogenesis.

The most detailed studies on cell proliferation have been made upon cultured cells, because under in vitro conditions many cell types display growth control very clearly. For example, when cultured fibroblasts are aliquoted into a tissue culture dish at low density, they readily attach to the dish surface and proceed to divide. Provided there is an adequate concentration of serum in the culture medium (see page 49) they then multiply until, as a flat monolayer, they become confluent. As soon as they cover the dish surface and reach its borders they stop dividing. Before confluence is reached, cell proliferation can be inhibited if the concentration of serum in the medium is significantly reduced (Figure 74). After reaching confluence, proliferation can be stimulated (but only into one or, occasionally, two further rounds of division) if the incubation medium is replenished with fresh serum.

There is a well-defined sequence of events in subconfluent cells undergoing repeated division; this sequence is called the 'cell cycle'. It should be noted that the physical process of subdivision (i.e. mitosis, when the divided chromosomes are segregated, and cytokinesis, when the parent cell separates in two) accounts for only a brief period of the cycle.

The cell cycle

Following a division at cytokinesis (D phase), there is a growth interval (G_1 phase) characterized by a high rate of DNA transcription and protein synthesis before DNA synthesis (S phase) begins (see Figure 75). When

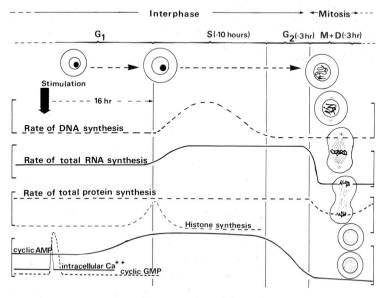

Figure 75. *Some of the events occurring during the cell cycle. Those indicated as taking place during the later stages of G_1 are based on studies in which mitotically quiescent cells have been stimulated to divide at the time indicated by the large arrow. The times and concentrations given should be taken only as a rough guide because there are wide variations between different systems.*

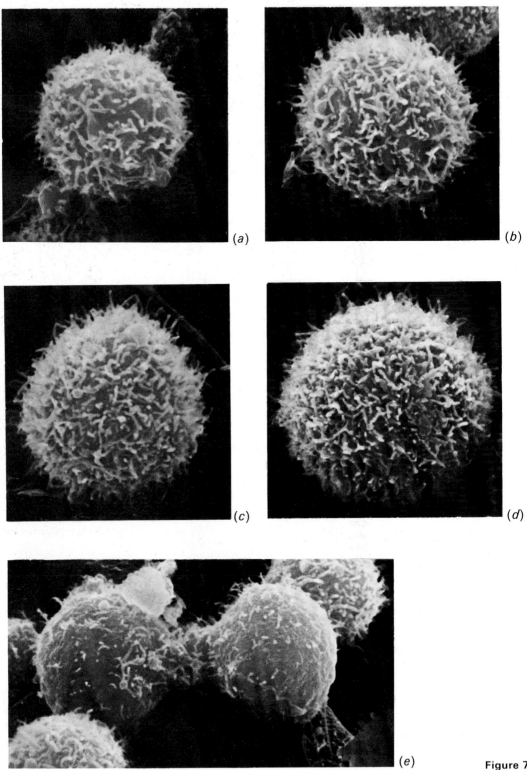

Figure 76

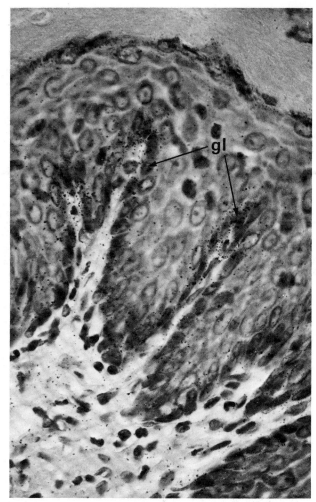

Figure 77. *Epidermal growth factor is a potent stimulator of cell division, but it affects only certain specific tissues. As its name suggests, one of these is the epidermis. This autoradiograph demonstrates that one of the reasons why the cells in the germinative (dividing) layer (gl) of this epithelium are responsive is because they have receptors that are able to bind the factor.*

To obtain this preparation, an animal was injected with radioactive growth factor, and four minutes later the tissue was taken and prepared for autoradiography. While most cells in and around the epidermal epithelium are unlabelled, those with receptors have a large number of grains over them.

Tissue culture studies indicate that within hours of binding the factor, the cells will enter S phase and divide. In man, neither the source nor the function of epidermal growth factor is known. Magnification ×1600.

Figure 76. *(Opposite.) When cultured cells divide, changes occur in their surface contour. The cells shown in this series of scanning electron micrographs have been cultured in suspension, so the changes observed are unconstrained by the effects due to attachment.*

In progressing from G_1 through to G_2 the cell increases in size and there is a significant increase in the number of surface microvilli. Compare Figures a and b (G_1) with c (S) and d (G_2).

At division (e), when daughter cells are produced, there is a dramatic increase in the surface to volume ratio and a marked reduction in the number of surface microvilli. Presumably the dense population of microvilli that accumulates by G_2 provides the cell with a reservoir of surface membrane that is drawn upon at division.

Magnifications: a to d ×5000; e ×4600.

Courtesy of S. Knutton and C. A. Pasternak, Department of Biochemistry, St George's Hospital Medical School, London.

the DNA has replicated there is a further brief period of growth (G_2 phase) before the prophase of mitosis begins. In an actively dividing preparation in culture (Figure 76), a full cycle normally takes about 24 hours; it is thus slower than that observed in vivo, where, as mentioned above, doubling times of as little as 8 hours may occur. During the cycle, G_1 usually (although it can be very variable) lasts about 10 hours.

When a quiescent preparation is stimulated to divide, either with serum or with specific growth factors (Figure 77), some cells take longer to respond than others, but there is always a characteristic 'lag phase' of several hours before DNA synthesis begins (Figure 75). It seems that this interval is necessary for the division stimulus to exert its required effects. Currently, much research interest is centred upon identifying the time when the beginning of the lag phase is reached in the G_1 phase of repeatedly dividing cells, since it is probably then that the cell becomes committed either to undergo another division or to stop dividing and enter interphase. This is a crucial switch-over point that, in normal tissues, determines when repeated division ceases and differentiation begins (Figure 78). In malignant cells it probably fails to operate.

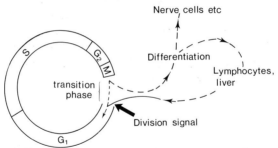

Figure 78. *The cell cycle and its relationship to differentiation. On reaching the postulated transition phase, the repeatedly dividing cells of the embryo progress directly, often without pausing, into the next division cycle. The same is true of stem cells (such as those of the bone marrow) in adult tissues. Eventually, however, some of the progeny of these dividing cells will cease to divide and they will 'differentiate' (i.e. take on their tissue-specific characteristics.)*
Some of these differentiated cells (such as lymphocytes) can, nevertheless, re-enter the division cycle if they are appropriately stimulated. When they do so, there is a lag of about 18 hours before DNA synthesis begins. After M. J. Berridge.

Studies on the arrest of division in culture and the effects of specific inducers of division (e.g. growth factors) are beginning to indicate the kinds of signal that are important in the control of cell division. Not unexpectedly, it seems that these signals are the same as those responsible for the control of other cellular processes like muscle contraction and secretion. In particular, it appears that cyclic nucleotides (such as cyclic 3',5'-adenosine monophosphate [cyclic AMP] and cyclic 3',5'-guanosine monophosphate [cyclic GMP]) and calcium ions are important. The interrelationships that exist between these messengers and the consequences of their actions are as yet unknown, but there is some indication of where their influence is probably decisive in the cell cycle. It is known, for example, that in fibroblasts at confluence the intracellular level of cyclic AMP increases. Moreover, the proliferation of malignant cells in culture is also inhibited when their intracellular levels of this cyclic nucleotide are experimentally increased.

Cyclic AMP

The regulation of gene expression at the level of transcription

During the development of the embryo, cell proliferation leads ultimately to differentiation; and when this occurs the characteristic, tissue-specific features of the different cell types appear. In amphibia, differentiated cell nuclei transplanted into enucleated oocytes have been shown to be capable of dividing and producing normal, fully differentiated adults. It can therefore be concluded that, like the nucleus in each fertilized egg, each differentiated (somatic) cell nucleus also contains a complete genome. The characteristic features of the different kinds of cell in the amphibian adult must therefore arise because only a proportion of the total gene complement is expressed. It is clear that, in the highly differentiated cells of mammals, most of the genome will be similarly repressed. The components primarily responsible for maintaining this permanent and extensive repressive regulation have not been identified, but obviously the most likely candidates are the nucleoproteins and nucleic acids, which, together with the DNA of the structural genes, make up the chromatin of the interphase nucleus.

Somatic cell

Genome

The histones

In the cell cycle the synthesis of histones has been shown to be coupled closely to DNA synthesis; this, together with structural alterations, such as the phosphorylation of histone H1 (see below), which occurs imme- diately before chromosome condensation in mitotic prophase, suggests that histones are involved in some way in nuclear regulation. However, because there are only five different kinds of histone molecule (desig- nated H1, H2a, H2b, H3 and H4), it seems unlikely that they have the range of structural variability required for the specific regulation of even small groups of genes. It seems more probable that they play a general, ancillary role. They may, for example, be important in the process of chromosome condensation during the early stages of mitosis.

Phosphorylation

Acidic, non-histone nucleoproteins

Unlike the histones, the acidic, non-histone proteins of chromatin dis- play considerable structural and functional variability. They vary in amount and form with changes in nuclear metabolism, and they may also become phosphorylated at specific times in the cell cycle. Even more significantly, their number and form is different in different cell types. Moreover, in reconstitution experiments using extracts of chromatin, it has been shown that a tissue-specific mRNA (for globin, which is found only in bone marrow) will be transcribed only in the presence of the acidic proteins derived from that tissue; those acidic proteins from other tissues cannot substitute for them.

Unfortunately, although there are these very promising hints, our understanding of the structure and function of the acidic, non-histone proteins is still very fragmentary and there is, as yet, no real indication of how they may be involved in the regulation of gene transcription.

(a)

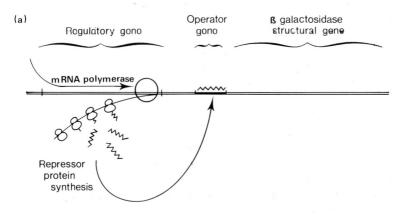

(b)

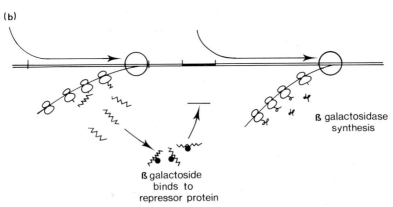

Figure 79. *Gene regulation in bacteria – the lac operon. The bacterium* Escherichia coli *is able to use the β galacto-side, lactose, as a source of carbon and energy only because it is able to manufacture the enzyme, β galactosidase (lac). However, because the galactoside regulates the transcription of the enzyme directly, the galactosidase is produced only when lactose is available.*

The regulation of β galactosidase synthesis may be itemized as follows:

1. β galactosidase is synthesized by the translation of an mRNA that is transcribed (by an mRNA polymerase – see Figure 71) on the β galactosidase structural gene.
2. Transcription on this structural gene is regulated by an immediately adjacent operator gene.
3. The activity of the operator gene is, in turn, controlled by an inhibitory 'repressor' protein that is synthesized on a regulatory gene.
4. Repressor protein bound to lactose is unable to bind and inhibit its operator. The presence of lactose thus allows ('induces') β galactosidase mRNA transcription and the synthesis of β galactosidase.

Nucleic acids

Although there is a large amount of information on the role of nucleic acids in the transcription of mRNA on eukaryotic structural genes, there are many other aspects of their activity that remain unexplained. Since it has not been possible in any simple manner to relate these activities to the transcription process itself, it seems likely that they may be of importance in its regulation.

In prokaryotes it is known that the major regulatory mechanism for the control of transcription on structural genes relies upon 'repressor proteins' made under the direction of 'regulatory genes'; this mechanism is described in the well-known 'operon model' of Jacob and Monod (see Figure 79). The repressor proteins are able to repress mRNA transcription by preventing the mRNA polymerase from gaining access to the structural gene. In eukaryotes, it is very likely that there is a similar hierarchical arrangement of genes, in which some genes regulate the activity of others. Since transcription and translation take place in different compartments of the cell in eukaryotes, it is conceivable that RNA rather than protein performs the role of repressor.

One observation that may be related to this hypothesis is that, in addition to the 'informative' sequences of eukaryotic DNA, a large amount (over 50 per cent) codes for repetitive sequences which have no meaning according to the genetic code. Some of the shorter sequences *Genetic code* probably play a 'structural' role, acting as 'spacers' (especially in the centromere regions), but the other, longer, sequences may well be con- *Centromere* cerned with regulation.

Another observation, which is of relevance to the idea that an RNA regulates transcription, is that although a large amount of RNA (known as 'heterogeneous' or 'polydisperse' RNA) is continuously transcribed in the nucleus, most of it is very short lived and only a small part of each molecule actually emerges into the cytoplasm as mRNA.

The regulation of gene expression in differentiated cells

In the regulation of transcription in bacterial systems there does not appear to be any process analogous to the cell differentiation of higher organisms. Instead, the entire genome is potentially available for transcription at all times, and differential gene expression depends upon a system which responds directly to the availability of metabolites in the surrounding environment (although the rate of production of some so-called 'constitutive' proteins is continuous and fixed). These 'flexible' *Constitutive* mechanisms of induction and repression that are able to control the syn- *proteins* thesis of proteins operate almost entirely at the level of transcription. They are the basis of the operon model referred to above, and they are now understood in considerable molecular detail.

In eukaryotic systems this kind of 'flexible' transcriptional control is much less evident. This is primarily because, in the differentiated cell, most of the genome is permanently repressed and cannot be readily induced. It can, perhaps, be regarded as a consequence of the division of labour that cell differentiation allows, since in higher organisms there is little variation in the chemical environment anyway. In man, only the intestine and liver are directly subjected to variations arising from gross fluctuations in diet, and it is the responsibility of the liver to moderate these effects by regulating the content of the blood.

Inductive regulatory mechanisms, operating primarily at the level of transcription, have nevertheless been identified in mammalian cells. Several, such as those involved in the detoxification of barbiturates (see page 154), are indeed related to the role of the liver in regulating blood content. Others, and these are the best documented, are a feature of the action of hormones such as the sex steroids and the thyroid hormones. These regulatory molecules are specifically concerned with the modulation of transcription in the fully differentiated cells of adult tissues.

The action of steroid sex hormones

'Target cell' see
Hormones

Many tissues of the reproductive system, and especially those of the uterus, respond to stimulation by sex steroid hormones such as oestrogen and progesterone. In general, these hormones stimulate an increase in protein synthesis within their target cells, which, in addition to providing for cell growth in general, includes the manufacture of characteristic (steroid-specific) proteins (Figure 80). The induced changes may take place rapidly; for example, within ten minutes of an intraperitoneal injection of oestrogen there is a detectable increase in mRNA synthesis in the uterus. For its maintenance, specific protein synthesis usually depends on continued stimulation by the steroid. In the

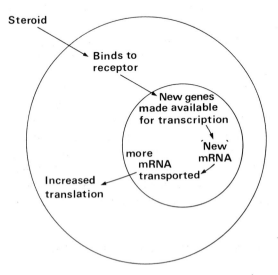

Figure 80. *The effect of a steroid hormone upon its target cell.*

best-documented system, the chick oviduct, it has been shown that chronic (18-day) stimulation with oestrogen provides each cell with an average of about 48 000 mRNA molecules concerned with directing the synthesis of the egg protein ovalbumin, but withdrawal of the hormone reduces the number to less than four per cell within ten days. A single injection of the hormone re-induces synthesis and, within little more than a day, the average number of ovalbumin mRNAs per cell reaches about 17 000.

Target cells for steroid hormones contain receptor molecules within their cytoplasm. These hormones gain access to the receptors because they are lipid soluble and thus readily able to diffuse across the primary permeability barrier of the plasma membrane. In studies in which differential centrifugation has been used to subfractionate uterine tissue, the receptor molecule for oestrogen has been identified in the high-speed supernatant (i.e. the cytoplasmic matrix) as a single protein with *Supernatant* a sedimentation coefficient of between 4 and 8 S. The receptor protein for progesterone is located in the same subcellular fraction, and has been identified as an ellipsoid-shaped molecule (mol. wt. 225 000) composed of two unequal subunits. These receptors bind to their respective hormones with a high degree of specificity.

The number of steroid receptors may vary considerably; in general, the greater the number of receptors the more responsive is the tissue. It has been shown, for instance, that chronic oestrogen stimulation can **Chronic.** Long (in addition to its other effects) increase the total population of progester- lasting. one receptors by twenty-fold, and with this increase there is a commensurate improvement in the response of the tissue to progesterone stimulation. Oestrogen withdrawal reduces the number of progesterone receptors dramatically.

Once the steroid binds to its receptor it becomes 'activated', and both molecules are then transferred, as a complex, to the nucleus, where they interact with the chromatin and stimulate the transcription of RNA. The activation — which occurs in the cytoplasm — is known to be temperature-dependent, but otherwise the nature and significance of this step is unclear. In the nucleus the steroid–receptor complex is known to stimulate mRNA synthesis on specific genes. Although it is known that acidic rather than histone nucleoproteins are concerned with this process, the molecular details of the interaction between the steroid–receptor complex and the chromatin remain to be defined.

In the early stages (i.e. the first few hours) of induction by steroid, the response is primarily a stimulation of mRNA synthesis, and it has been shown directly that the number of initiation sites for mRNA transcription increases. Later the synthesis of other RNA species, and especially those required for the production of ribosomes, is stepped up (see below).

Differential gene expression in relation to gene amplification

Given the availability of a structural gene for transcription, the speed at which it can be transcribed becomes a major rate-limiting step in the pathway that leads to protein synthesis. This constraint can be overcome if the number of structural genes available for any particular translation product is increased. This arrangement, known as 'gene amplification', does exist. It is known, for example, that in all cells there are multiple (up to 500) copies of genes concerned with directing histone synthesis. Amplification in this instance may be necessary to cater for those brief periods early in the S phase when a high rate of histone synthesis is required to keep pace with the rate of DNA replication (see Figure 75). Similarly, there are multiple (up to 1000) copies of the structural genes required for the transcription of ribosomal and transfer RNAs. This arrangement makes up for the fact that these molecules represent the end-product of a single synthetic step, transcription. Unlike proteins, whose numbers are increased at a later stage by the repeated translation of a single mRNA (see page 137), there can be no opportunity for an increase in the number of RNAs at a post-transcriptional level.

Haploid

'Globin chains' see *Haemoglobin*

For most, if not all, other structural genes, however, the available evidence suggests that there is only a single copy in each haploid complement of chromosomes. Thus, even in situations like the developing erythrocyte, where the cell is almost exclusively concerned with the intensive manufacture of only two translation products (more than 90 per cent of the total proteins synthesized are α- and β-globin chains), translation depends entirely upon the mRNA transcribed on single copies of their structural genes.

It seems, therefore, that gene amplification occurs for only those basic products that are required in bulk in all kinds of cell. For most of the proteins specific to particular differentiated cells, mRNA is transcribed on a single gene; the amplification of this information occurs only: (a) by the repeated transcription of this gene, and (b) by the repeated translation of mRNA (see below).

Transcription of mRNA and the regulation of gene expression at the post-transcriptional level

Unlike the prokaryotic system, in which protein synthesis is coupled tightly to transcription (see Figure 71), the synthesis of protein in eukaryotes is carried out in a discrete cellular compartment (the cytoplasm) and it is thus well separated from transcription. This arrangement requires the mRNAs to be longer lived, and it provides an additional step at which to regulate the information derived by gene transcription. In eukaryotes, therefore, although the regulation of gene expression occurs primarily at the level of transcription, post-transcriptional control is also possible. Indeed, it is probably very important.

As in bacteria, mammalian mRNAs are synthesized by polymerases using the nucleotide sequences of DNA as the template. mRNAs able to direct the synthesis of proteins such as the globin polypeptides of haemoglobin, myosin and immunoglobulin have been isolated and puri-

'Immunoglobulin' see *Antibodies*

fied and their structure elucidated. The fidelity of these purified messengers has been established by their ability to direct the synthesis of their particular protein in artificial, 'cell-free' systems.   *Cell-free systems*

Pre-mRNA

The RNA molecules initially transcribed on structural genes in eukaryotes (the 'primary gene products') are considerably (about 10-fold) larger than those required to carry the necessary coded information for protein synthesis. These molecules are most appropriately called 'pre-mRNAs', because, almost as soon as they arise, most of their 'non-informative' nucleotide sequences are removed by nuclear enzymes. The functional importance of these sequences in the regulation of gene expression is at the moment unknown.

Control of the transport of mRNA to the cytoplasm

There is good reason to believe that an important step in the regulation of gene expression at the post-transcriptional level is the selection of which mRNA species are to be transported to the cytoplasm. Unfortunately, however, the information available on these processes is fragmentary. It has, nevertheless, been shown that normally only a small proportion of the mRNA arising from pre-mRNA in the nucleus actually reaches the cytoplasm, and it is also apparent that in slowly growing cells, where there is less protein synthesis, there is proportionally less nuclear mRNA transported. From other studies it appears that the amount of mRNA reaching the cytoplasm is dramatically reduced unless a functional nucleolus is present in a nucleus, and it therefore seems possible that the nucleolus may be implicated in processing mRNA for transport.

No information is available on the molecular modification that allows an mRNA to be selected for transport to the cytoplasm, although there is evidence that some mRNAs may become associated with certain, as yet unidentified, proteins while they are in the nucleoplasm. Conceivably these proteins are able to identify and bestow protection upon the small proportion of mRNA species that are destined for transport to the cytoplasm.

The addition of polyadenine nucleotides

Another kind of structural processing that concerns mRNAs within the nucleus is the addition of a length of about two hundred residues of the nucleotide adenine ('poly A'). This addition is made to most mRNAs, although for some unknown reason those directing the synthesis of histones are excepted. Since the poly A remains attached to a large proportion of the mRNAs that reach the cytoplasm, its presence is thought to be of primary importance to the events taking place outside the nucleus (see page 135).

The density and staining of chromatin and its relation to transcription

Heterochromatin and euchromatin

Although the specific agencies concerned with controlling gene expression are unclear, the morphological correlate of gene activation, in the form of gross microscopic changes in the appearance of nuclear chromatin, is readily observed. In most cells the chromatin network of the interphase nucleus is variously condensed and expanded, and in some cell types it may have a very characteristic distribution. The condensed regions stain intensely with basic dyes and are said to contain 'heterochromatin'. A predominance of heterochromatin is found typically in cells with inactive nuclei. A good example is found in the development of the erythrocyte (Figures 81 and 82) where, in the final stage (the normoblast), the nucleus has effectively 'retired' and is about to be ejected from the cell. Another example is found in the small lymphocyte, where the nucleus, although relatively large and long lived, remains quiescent (unless called upon to participate in the immune response) throughout its lifespan (Figure 83).

Lymphocyte

By comparison, expanded chromatin regions stain much less intensely and are said to contain 'euchromatin'. Euchromatic nuclei are typical of metabolically 'active' cell types, such as the liver parenchyma cell (see Figure 69) and the ventral horn neurone, but it is important to note that the amount of euchromatin in a nucleus reflects the number of genes being transcribed rather than the amount of mRNA produced (Figures 82 and 83). In the plasma cell, for example (see Figure 97), where there is a high rate of protein synthesis, the nucleus is relatively heterochromatic. This is because these cells are concerned with making large amounts of a single kind of translation product (the immunoglobulin that they secrete). Their requirement for mRNA is thus very narrow and specific.

Plasma cell

'Immunoglobulin' see Antibodies

Autoradiographic studies support the idea that mRNA is transcribed in euchromatic regions, since they show that RNA precursors are initially incorporated within them. Indeed, electron microscope studies using autoradiography in association with thin sections specifically stained for RNA have identified fibrils at the boundary between heterochromatin and euchromatin that are believed to represent the primary gene product, pre-mRNA (Figure 84).

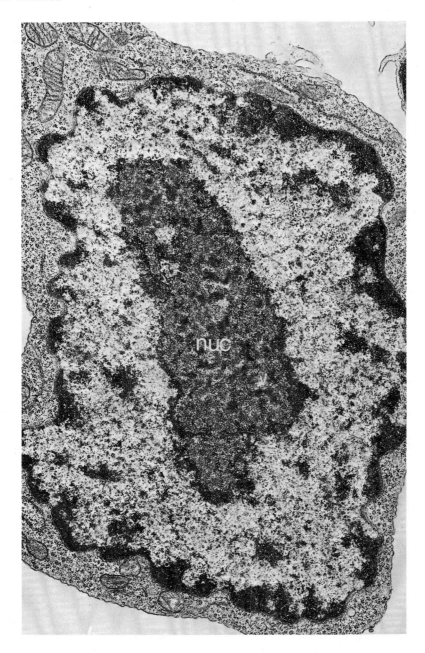

Figure 81. *The nucleus of an erythroblast at an early ('proerythroblast') stage of development, when there is a high rate of transcription and a high rate of RNA transport to the
cytoplasm. Characteristically, heterochromatin is restricted to the periphery, below the
nuclear membrane, while the central nucleoplasm is dominated by a large nucleolus (nuc).
Magnification ×14 000.*

Figure 82. *A late stage in erythrocyte development ('the normoblast'). Transcription has ceased and haemoglobin synthesis in the cytoplasm depends upon mRNA and ribosomes transcribed during the earlier stages. The nucleus is now much smaller in volume (compare magnification with Figure 81) and there is a much greater proportion of heterochromatin. Magnification ×26500.*

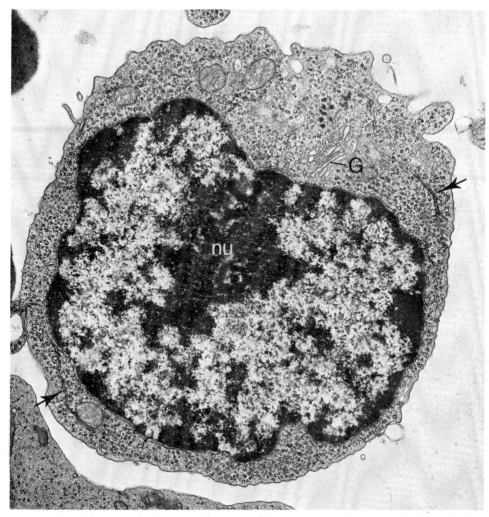

Figure 83. *A recently divided lymphocyte in bone marrow. The nucleus is relatively large and contains a well-developed nucleolus (nu). Within the cytoplasm is a dense population of free polyribosomes concerned with the synthesis of endogenous structural proteins. Single strands of rough endoplasmic reticulum are indicated (arrows). Compare with Figure 97. Golgi complex – G. Magnification ×18 000.*

Cell fusion and the reactivation of heterochromatic nuclei

An alternative approach that also allows changes in nuclear metabolism to be related directly to changes in nuclear morphology is to use the cell fusion method to reactivate inactive, heterochromatic nuclei. In these studies, a cell surface-active animal virus (killed by irradiation with ultraviolet light) is used to fuse different cell types from different species in vitro. The hybrid cells produced are known as 'heterokaryons'. In a heterokaryon produced by the fusion of avian (chicken) erythrocytes

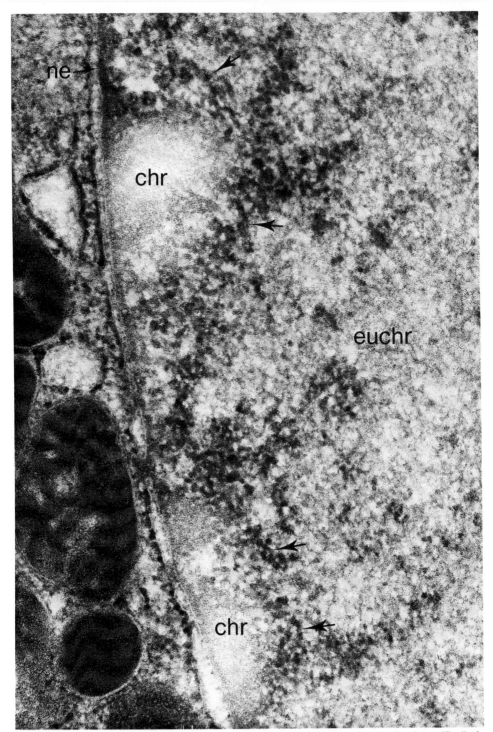

Figure 84. *An electron micrograph of the periphery of a rat liver cell nucleus, stained specifically for RNA. This technique bleaches the heterochromatin (chr) and demonstrates the distribution of perichromatin RNA fibrils (arrowed) at the periphery of the euchromatin (euchr). These fibrils are believed to represent newly formed premRNA.*

    Note also the intense staining of the RNA in the ribosomes on the cytoplasmic surface of the nuclear envelope (ne). Magnification ×73 000.

    Courtesy of G. Moyne, Institut de Recherches Scientifiques sur le Cancer, Villejuif, France.

(which have heterochromatic nuclei and are normally unable to incorporate DNA or RNA precursors) with cultured mammalian cell types
(in which the nuclei are active in both transcription and DNA replication), the avian nuclei become reactivated and are able to incorporate
both RNA and DNA precursors. As they become reactivated, the avian
nuclei show well-defined morphological changes (Figures 85 and 86).
To begin with, they show a dramatic increase in volume and mass and
their tightly condensed heterochromatin becomes dispersed. RNA synthesis then begins, and, in a day or so after fusion, DNA replication
occurs and nucleoli appear. Once nucleoli are present, the characteristic
avian proteins begin to be synthesized in the cytoplasm. Since all the
ribosomes in the heterokaryon are derived from the mammalian cell
(there are none to begin with in the avian erythrocyte) these proteins
must be synthesized in response to new mRNA arising in the reactivated
and now euchromatic avian nucleus.

These experiments are also of interest in the context of gene regulation. They show, for example, that major changes in nuclear activity,
including both DNA replication and transcription, can occur in response
to signals from the cytoplasm. In addition since, in this instance, an avian
nucleus is seen to respond to a mammalian cytoplasmic environment,
they also suggest that these signals are of a general rather than a specific
nature.

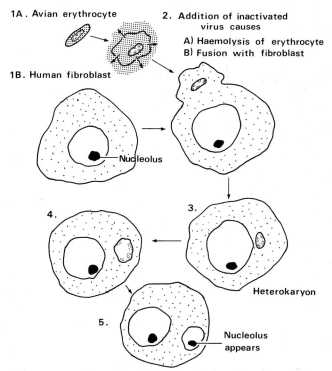

Figure 85. *An outline of the sequence of events leading to reactivation of the avian erythrocyte nucleus in a heterokaryon.*

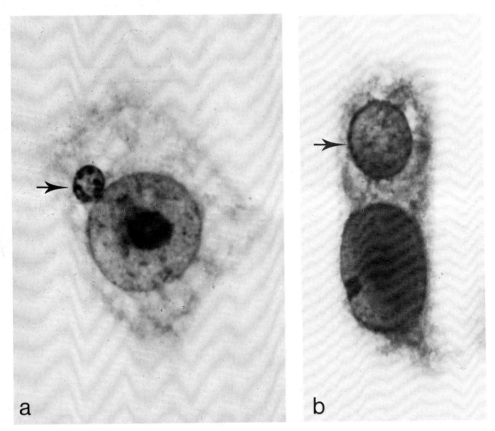

Figure 86. *Heterokaryons in which a cultured human cell contains a hen erythrocyte nucleus. In (a), the heterokaryon is newly formed, and contains an erythrocyte nucleus (arrow) that is about the same size and shows the same dense, patchy concentrations of heterochromatin as the donor cell nucleus. In (b), with time, the erythrocyte nucleus (arrow) has enlarged and has become entirely euchromatic. The morphology of the nucleus of the recipient cell appears to have altered in (b), but in fact its form does not change. Note that the magnifications of both micrographs are the same. (×3200).*
   Courtesy of H. Harris, Sir William Dunn School of Pathology, Oxford.

THE FINE STRUCTURE AND FUNCTION OF THE NUCLEOLUS

Nucleoli are concentrations of densely staining, basophilic material that appear to lie free within the nucleoplasm of the interphase nucleus (Figure 87). They vary in number in different cell types, but their number and distribution are usually characteristic for any given cell type. They are rich in ribonucleoprotein and are intimately associated with the specific region of the chromosomal DNA that codes for ribosomal RNA. As seen in the electron microscope, nucleoli consist of a mass of peripheral granules (15 to 20 nm diameter) embedded within an amorphous proteinaceous matrix and surrounding a central area containing short (10 nm diameter) fibres (Figure 88).

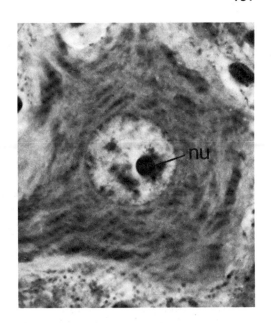

Figure 87. *A light micrograph of a nerve cell body in the spinal cord. The large euchromatic nucleus contains a prominent spherical nucleolus (nu). The clumps of basophilic material in the surrounding cytoplasm are typical features of the nerve cell body and are known as 'Nissl bodies' (see Figure 109). Magnification ×400.*

Within the nucleolar area the DNA which contains multiple copies of the genes for ribosomal RNA is transcribed. This transcription is the responsibility of a specific kind of RNA polymerase found only in the nucleolus. The transcribed RNA is then processed so that the initially formed primary gene product, a long, 45 S precursor RNA filament, is degraded into shorter lengths to yield (amongst its multiple products) a 32 S and an 18 S component. The 32 S components become further degraded to 28 S products, which, within the nucleolus, associate with their respective ribosomal proteins to form the large ribosomal subunits. The 18 S components also associate with ribosomal proteins within the nucleus to form the small ribosomal subunits (see below), but it has not been established if this occurs within the nucleolus or the nucleoplasm. It is clear, however, that the final assembly of discrete ribosomes from the two complete subunits occurs only when the various components reach the cytoplasm (Figure 89).

By using selective enzyme digestion and selective staining, the nucleolar organizer region of the chromosome can be identified within the nucleolus. Otherwise, however, although pulse/chase autoradiography suggests that labelled ribosomal RNA precursors move first to the central fibrous elements and later to the peripheral granules, it is difficult to correlate the fine structure of the nucleolus with its role in processing ribosomal RNA.

As with most aspects of eukaryotic nuclear metabolism, nucleolar functions thus remain to be fully evaluated. It is clear that the production of ribosomal RNA is a major responsibility, but other important roles, such as the processing of mRNA, mentioned above, have yet to be clearly defined.

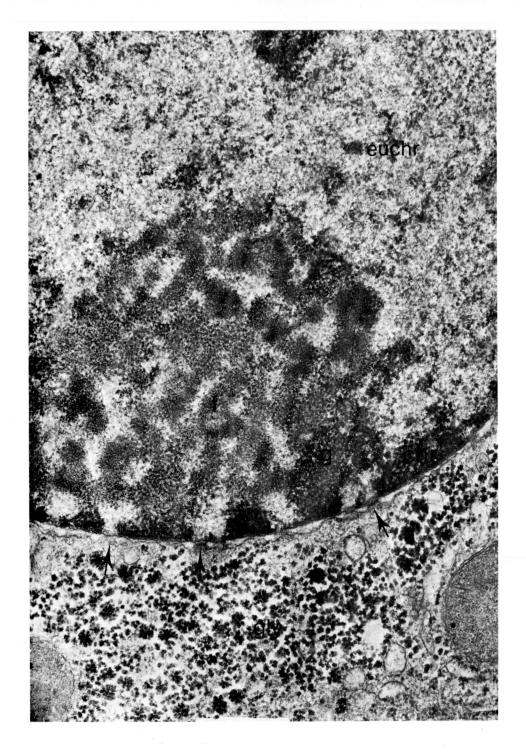

Figure 88. *A portion of the nucleus in a liver parenchyma cell. The nucleolar region, lying within the euchromatin (euchr), consists of a mass of granular (g) and fibrillar (f) components. Immediately adjacent to this region, and below the nuclear membrane, are dense aggregations of chromatin (chr), the distribution of which alternates with that of the nuclear pores (arrowed). In the cytoplasm are coarse aggregates of glycogen (gly) (see page 207). Magnification ×43 000.*

Courtesy of G. Moyne, Institut de Recherches Scientifiques sur le Cancer, Villejuif, France.

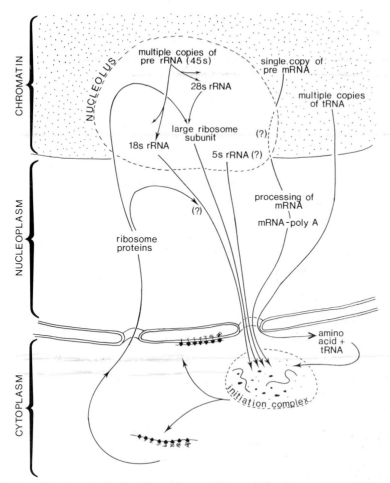

CHROMATIN

NUCLEOPLASM

CYTOPLASM

NUCLEOLUS

multiple copies of
pre rRNA (45s)

single copy of
pre mRNA

28s rRNA

multiple copies
of tRNA

large ribosome
subunit

(?)

18s rRNA

5s rRNA (?)

(?)

processing of
mRNA

mRNA-poly A

ribosome
proteins

amino
acid +
tRNA

initiation complex

Figure 89. *An outline showing the locations where post-transcriptional processing of RNA takes place.*

THE NUCLEAR ENVELOPE

The nucleus is surrounded by a cisternal element, the nuclear envelope. This flattened, sac-like compartment provides a continuous boundary between the nucleoplasm and the cytoplasm, except where it is penetrated by characteristic pores (Figures 69 and 90). The envelope has much in common with the cisternae of the rough endoplasmic reticulum, and often bears ribosomes on its cytoplasmic surface. Occasionally, there is continuity between the membranes of the nuclear envelope and the rough endoplasmic reticulum, and both kinds of cisternal element may contain, within their lumina, secretory products produced by their attached ribosomes (see Figure 100).

Cisternae.
Membrane-limited, closed compartments.

Figure 90. *The surface of the nuclear envelope seen in a freeze-fractured replica of a kidney cell. The distribution of the nuclear pores is clearly displayed. At the margins of the nucleus, the freeze-fracture plane runs through the outer (cytoplasmic) cisternal membrane; then it jumps, forming a ragged border (arrows), and runs through the inner (nucleoplasmic) cisternal membrane. The step at the border gives an idea of the thickness of the envelope. Magnification ×18 500.*
  Courtesy of L. Orci and A. Perrelet, Institute of Histology and Embryology, University of Geneva.

Nuclear pores may occupy as much as 20 per cent of the nuclear envelope. Each pore consists of a circular or sometimes octagonal opening (40 to 100 nm diameter) penetrated by a so-called 'pore complex'. High-resolution studies suggest that the pore complex itself consists of a cylinder made up of eight regular subunits arranged about a dense

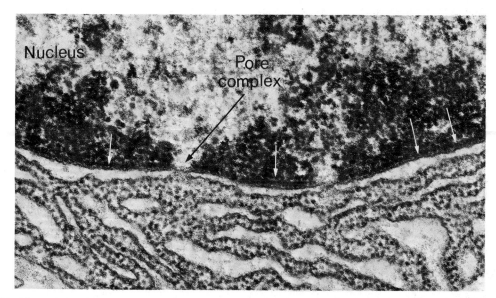

Figure 91. *The periphery of the nucleus in a cell from the exocrine pancreas. The arrows indicate the fibrous lamina at the boundary of the nucleoplasm beneath the nuclear envelope. The pore complex that inserts into the aperture in the nuclear envelope is an integral part of the lamina. Here it is represented only as a thin fuzzy line. Magnification ×40 000.*

central granule; each complex appears to be an integral part of the granular to fibrous lamina that covers the nuclear surface and lies immediately beneath the nuclear envelope (Figure 91). The precise functional significance of the lamina and its component pore complexes is obscure, although it seems likely that the lamina provides attachment sites for chromosomes and may be important for their arrangement.

Clearly, the pore complexes play some role in the transfer of information between the nucleus and the cytoplasm, but at present the available evidence indicates only that the permeability of nuclear pores varies with the functional state of the cell. Using tracers introduced into the cytoplasm, it has been shown that optimally the pores are permeable to molecules with a maximum diameter of only 4.5 nm.

The number of pores per unit area of nuclear surface also varies with changes in nuclear activity. For example, in the lymphocyte it appears that the number of nuclear pores increases soon after nuclear activity is stimulated. In the maturing erythrocyte, on the other hand, when the increasingly heterochromatic nucleus is about to be expelled from the cell, the number of nuclear pores decreases very rapidly.

PROTEIN SYNTHESIS – CYTOPLASMIC EVENTS

Although the nuclear envelope confines transcription and the earlier stages of mRNA processing to the nuclear compartment and restricts translation to the cytoplasm, protein synthesis is, nevertheless, most conveniently discussed as a single, sequential process. We will now, therefore, outline the events that take place in the cytoplasm between the ribonucleic acids and the ribonucleoprotein components that arise in the nucleus. These events comprise the process of translation, and in the eukaryotic system they take place at a specific interface in the cytoplasmic matrix — the surface of the ribosome.

Ribosome structure

Ribosomes are small particles with a diameter of between 17 and 23 nm. They contain RNA and protein in about equal amounts, and have a subunit structure that requires magnesium ions for its integrity. Thus, when magnesium ions are removed from ribosomal or microsomal preparations (see page 53), the individual ribosomes, which in mammals have a characteristic sedimentation coefficient of 80 S, split into a large *S unit* (60 S) and a small (40 S) subunit. The larger, 60 S subunit contains a 28 S RNA and a 5 S RNA, while the smaller, 40 S subunit contains *Subunit* a single 18 S RNA molecule. The 5 S RNA, like the other RNA species, arises in the nucleus by transcription of DNA. The distribution of these rRNAs within the ribosomal subunits is not yet clear, but it is probable that they are exposed over large parts of the ribosome surface. They are, for example, susceptible to ribonuclease attack, and it is because their phosphate groups are able to bind basic dyes that dense populations of ribosomes in the cytoplasm give rise to 'basophilia'. Their full functional significance is not yet clear either, although it is known that they play an important role in providing the specific binding site for mRNA attachment.

In mammalian ribosomes the number and distribution of ribosomal proteins remain to be clearly elucidated, but in bacterial ribosomes, which are in many respects very similar, the small subunit contains 21, and the large subunit probably more than 30, different proteins. Some indication of the functional importance of these proteins is given by studies concerned with experimentally reassembling bacterial ribosomes from their constituent RNA and protein molecules. In these studies it has been shown that ribosomes susceptible to inactivation by the antibiotic, streptomycin (an inhibitor of protein synthesis), can be made resistant by first dismantling them and then, in their reassembly, replacing a single protein in the small subunit with its counterpart derived from the ribosomal subunits of a streptomycin-resistant strain.

One of the proteins already identified in the large ribosomal subunit is peptidyl transferase, the enzyme responsible for making the peptide bond between the amino acids in the growing polypeptide chain.

Translation

The molecular events concerned with the directed synthesis of polypeptides at the ribosome are again known largely from studies of pro-

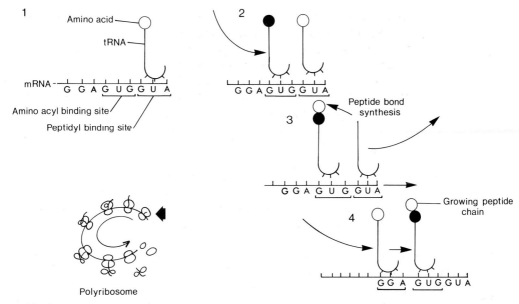

Figure 92. *An outline of translation in prokaryotes. Translation is initiated when a codon near the terminus of the mRNA nucleotide sequence identifies the binding site on the small ribosomal subunit (provided by bases in the 18 S rRNA). Thereafter, the mRNA is aligned so that its codons (triplets of nucleotide bases) can, in turn, dictate which tRNA molecules bind to a specific site (the amino acyl site) on the ribosome (see Figure 93). As indicated in (1), there are in fact two tRNA binding sites on each ribosomal subunit, the amino acyl binding site and the peptidyl binding site. In prokaryotes the first codon in the mRNA dictates that the first amino acid carried to the ribosome by its tRNA is always N-formyl-methionine; as this acyl–tRNA is transferred to the peptidyl site, the large ribosomal subunit attaches to the small subunit. The initiation complex is then complete and the amino acyl site is available for binding a tRNA carrying the first designated amino acid of the polypeptide chain (2). When this acyl–tRNA binds, peptidyl synthetase, an enzyme component of the large subunit, links the two adjacent amino acyl groups via a peptide bond and the formyl-methionine is translocated from its carrier tRNA to the amino acyl–tRNA occupying the amino acyl binding site. (3) The deacylated formyl-methionine–tRNA next vacates the peptidyl binding site and the mRNA moves on, bringing the tRNA carrying the growing chain of amino acids into position at the peptidyl binding site. The amino acyl binding site thus becomes vacant again and is able to bind the next designated amino acyl–tRNA (4). This series of steps is repeated until a codon signalling the termination of the polypeptide chain reaches the peptidyl binding site.*
G – guanine; A – adenine; U – uracil.

karyotic systems. However, since the fine molecular details of this process are not strictly within the scope of this book, they are given only in outline, in Figures 92 and 93.

The process of translation in the eukaryote appears to be essentially the same as that of the prokaryote, although there are minor differences (for example, the initiating amino acid appears to be methionine and not formyl-methionine). As an amplification step in the use of the transcribed information, however, translation is a more significant process in eukaryotes. This is because, by comparison with its prokaryotic counterpart (page 122), each mRNA molecule is relatively long lived and can be translated and retranslated many times. In cultured fibroblasts, for example, it has been estimated that the mRNA population turns over only about once per cell cycle. The factor limiting the number

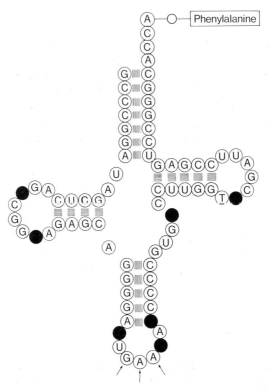

Figure 93. *Transfer RNA (tRNA). These single-stranded RNA molecules are synthesized on chromosomal DNA templates and are transported to the cytoplasm, where they bind to their respective amino acid residues. The cytoplasmic matrix thus contains a pool of amino acyl–tRNA complexes available for use in translation.*

When it is formed, each acyl–tRNA complex has a high-energy bond incorporated into it, and during the process of translation this energy becomes available for the synthesis of the linkage between each successive amino acid.

All tRNAs thus far examined have certain stretches of their nucleotide sequences in common, and thus, the nucleotide chain in all of them probably folds in the same 'clover leaf' manner shown here. The triplet group ACC, which carries the amino acid, is always found at the 3′ terminus, while the triplet group (indicated by the arrows) on the third arm-like extension represents the 'anticodon'. In translation it is the anticodon that identifies and binds to the template provided by the codons of the mRNA (see Figure 92).

The tRNA shown here is the phenylalanine–tRNA of the bacterium Escherichia coli.

G – guanine; C – cytosine; A – adenine; U – uracil.

of times an mRNA can be translated is not known, but there is reason to believe that it may be related to a gradual reduction in the length of the attached poly A.

In the eukaryotic system, translation also provides an opportunity for the further regulation of gene expression. In the oocytes of some invertebrate species, for example, it has been shown that increased protein synthesis can be initiated in the cytoplasm by an extracellular signal (fertilization) 'unmasking' mRNAs that, until that time, have been present but unavailable for translation. To what extent this and other mechanisms allowing control at the translation level play an important role in modulating protein synthesis in eukaryotic cells in general is, however, not known.

Polyribosomes

In eukaryotic cells as in prokaryotes (see Figure 71) mRNA molecules are processed by several ribosomes at once and these multi-ribosome–mRNA complexes are called 'polyribosomes' (or 'polysomes') (Figures 92 and 95). The number of ribosomes in a polyribosome is a function of the length of the mRNA molecule with which it is associated. Thus, for example, in a polyribosome concerned with the synthesis of a very long polypeptide chain like that of serum albumin (mol. wt. 64 000), the mRNA will be long enough to carry as many as 20 constituent ribosomes. As each ribosome translates the mRNA, its associated polypeptide chain becomes progressively longer so that, in a polyribosome at any given time, the constituent ribosomes will be associated with varying lengths of polypeptide depending upon their position (Figure 92). Polyribosomes obviously provide an added dimension to the economy with which the mRNA information can be handled, since the only limited factor in this arrangement is the speed at which each ribosome can translate the message.

Serum albumin. The major blood protein; it accounts for 50 per cent of all plasma proteins.

Membrane-bound polyribosomes

Polyribosomes that lie free within the cytoplasmic matrix release their formed polypeptides into the matrix. The destination of these polypeptides varies; some, such as histones and ribosomal proteins, are transported into the nucleus, while others, such as the contractile proteins, actin and myosin, are directed to their preferred location within the cytoplasm itself. The agency that directs the transport of these molecules is unclear. Other polyribosomes, however, and most notably those concerned with synthesizing proteins destined either to become integral components of membranes or otherwise resident within cisternal compartments (for example, the contents of secretory granules) are predominantly attached to cisternal membranes. Cisternae bearing polyribosomes typify the rough endoplasmic reticulum, a prominent feature in growing cells that reaches its most dramatic development in cells synthesizing copious amounts of protein secretion (Figures 94 and 95). Membrane-bound polyribosomes are attached to the cytoplasmic surface of cisternal membranes by the large subunits of their constituent ribosomes.

Translation on polyribosomes attached to rough endoplasmic reticulum cisternae

As with free polyribosomes, translation begins when the initiation complex comes together in the cytoplasmic matrix. Then, as shown in Figure 96, the attachment of the polyribosome to a cisternal membrane occurs. Attachment is almost certainly determined by the synthesis of a signal sequence at the beginning of the emerging polypeptide chain. The signal

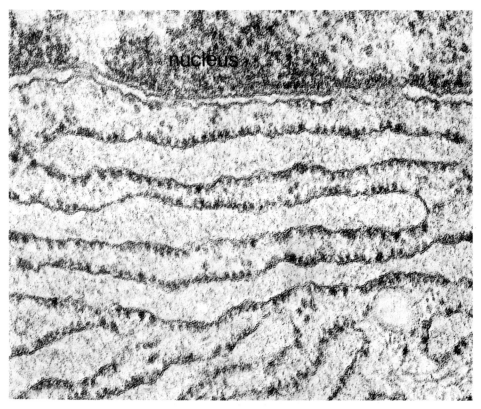

Figure 94. *Rough endoplasmic reticulum cisternae in an acinar cell of the exocrine pancreas. The packing of the polyribosomes on the cytoplasmic surface of these cisternae is extremely close and probably approaches maximum capacity. Polyribosomes also occur on the cytoplasmic surface of the nuclear envelope, but in fewer numbers. Magnification ×60000.*

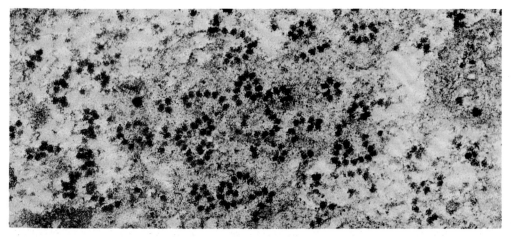

Figure 95. *A grazing section across the cisternal surface of the rough endoplasmic reticulum, showing the distribution and form of the attached polyribosomes.*
   Courtesy of M. Steer, Department of Botany, Queen's University, Belfast.

peptide is made under the direction of the attached mRNA and consists of about twenty amino acids. Where several different polypeptide chains are synthesized within a particular cell type, it is probable that they all carry the same signal sequence. Reconstruction experiments have been carried out in which mRNA for secretory proteins (which bear the code for the signal terminus) and rough microsome membranes (rough endo-plasmic reticulum cisternae stripped of their attached ribosomes) from pancreatic cells have been mixed with free ribosomes from a variety of sources. In these experiments the polyribosomes form and attach to the membranes, demonstrating directly that it is the signal sequence alone that is responsible for inducing attachment.

Translation of the mRNA is essentially the same as that which occurs on free polyribosomes. The destination of the growing polypeptide chain is, however, different, since it is always transferred across the membrane and into the cisternal lumen. The manner in which the large ribosomal subunit attaches to the membrane is not known, but it may be a primary function of the signal sequence to cause an aggregation of suitable receptor binding proteins on the cytoplasmic surface of the cisternal membrane.

The insertion of the newly synthesized protein into and through the lipid bilayer of the cisternal membrane is also an essential requirement in this kind of translation process (Figure 96). In this regard, the nature of the initial signal sequence may be important, since it is a characteristic of these sequences that they include several hydrophobic residues. These residues can be expected to facilitate the penetration of the signal peptide through the lipid bilayer.

In experimental systems in which mRNAs normally translated on membrane-bound polyribosomes are translated on free polyribosomes (i.e. without microsomal membranes), the signal peptide remains attached to the newly synthesized polypeptide and is not removed.

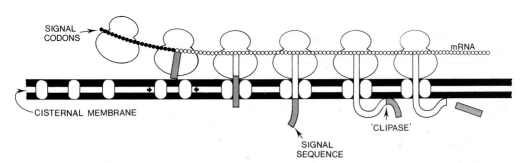

Figure 96. *Translation on an attached polyribosome – the signal hypothesis.*
The signal codons on the mRNA are known to direct the synthesis of a signal sequence at the beginning of the growing polypeptide chain. The primary purpose of this sequence is to induce the ribosome-mRNA initiation complex to identify and attach to the cisternal membrane of the rough endoplasmic reticulum. Attachment probably then induces an aggregation of proteins within the cisternal membrane, which facilitates the directed transfer of the growing polypeptide into the cisternal lumen. The early removal of the signal peptide is presumably effected by a membrane-bound enzyme – the 'clipase'. After G. Blobel.

When, on the other hand, polypeptides are transported into the rough endoplasmic reticulum, the signal sequence is soon lost. It seems, therefore, that once the growing polypeptide chain becomes long enough for the signal peptide to become exposed on the luminal side of the cisternal membrane, the signal region is rapidly removed by a membrane-associated peptidase (a 'clipase').

6

INTRACELLULAR COMPARTMENTS

The cisternae of the rough endoplasmic reticulum provide an enclosed intracellular compartment within which newly synthesized products can be segregated. The cisternal membranes that enclose this compartment are continuous and their general physical characteristics, especially as far as their impermeability towards macromolecules is concerned, are similar to those of the plasma membrane (see page 85). The capacity to concentrate and separate the processing of some of their constituents within closed, membrane-bound compartments is a feature of all eukaryotic cells. This, together with the advantage of the additional membrane interface it provides (which itself adds an extra dimension to the complexity of the organization), has allowed an elaborate division of labour and has enabled the cells of eukaryotes to obtain such a high order of functional complexity.

Other intracellular compartments, which similarly provide for an additional specialization of function, include: cisternal systems such as the smooth endoplasmic reticulum and Golgi complex; vesicular elements such as lysosomes, peroxisomes and secretory granules; and discrete, complex organelles, like mitochondria. Although all of these *Organelle* compartments are functionally interrelated, each of them has a characteristic morphology and distribution. For convenience they will be described separately.

THE ROUGH ENDOPLASMIC RETICULUM

Structure

The cisternae of the rough endoplasmic reticulum, by definition, bear polyribosomes on their cytoplasmic surface. They are virtually ubiquitous cell components that reach their most dramatic development in protein-secreting cells (Figure 97).

The forms of the cisternae are variable. In some cell types, as their name suggests, the cisternae constitute a reticulum of tubular elements (e.g. the fibroblast – Figure 98). In others, where the demand for protein synthesis is greater, they form wide, flattened sacs arranged in a compact, parallel array (as in the acinar cells of the exocrine pancreas –

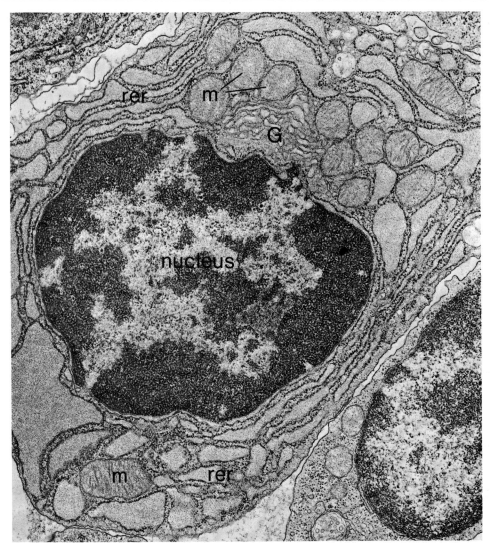

Figure 97. *A plasma cell. This kind of cell synthesizes and secretes copious amounts of antibody (immunoglobulin). Since antibodies are glycoproteins, the cell is characterized by a well-developed (and often distended) rough endoplasmic reticulum (rer) and a prominent Golgi complex (G). The antibody is released as soon as it is synthesized, and there thus appears to be no requirement for storage (secretory) granules. The nucleus of the plasma cell is moderately heterochromatic (see page 124); and typically, the condensed chromatin is distributed as thick, peripheral blocks in a radial ('clock-face') arrangement.*

At lower right is a developing erythroblast. The dense population of free ribosomes, which is concerned with the synthesis of haemoglobin (for 'home' consumption), contrasts markedly with the rough endoplasmic reticulum of the plasma cell, which is synthesizing protein for 'export'.

Figure 98. *(Opposite.) A micrograph, taken with a high-voltage electron microscope, showing the edge of a tissue culture cell. The improved penetrating power of the electrons, gained by using a greatly increased accelerating voltage, allows the full thickness of the cell periphery to be examined in detail. Within the cytoplasm there is a widely distributed network, the endoplasmic reticulum (arrows). A cluster of dense filamentous mitochondria are labelled (mit), while elsewhere in the cytoplasm there are several clearly defined bundles of microfilaments (mf) or 'stress fibres' (see page 223). At the cell boundary there are spiky microvilli (mv), which also appear to contain a core of microfilaments.*

On the left-hand margin of the micrograph the cell increases in thickness and the electron beam fails to penetrate. The sharply defined opaque lines projecting into the cell at lower left represent fine creases in the preparation; these cause similar penetration difficulties.

Courtesy of I. K. Buckley, Department of Experimental Pathology, John Curtin School of Medical Research, Canberra.

144

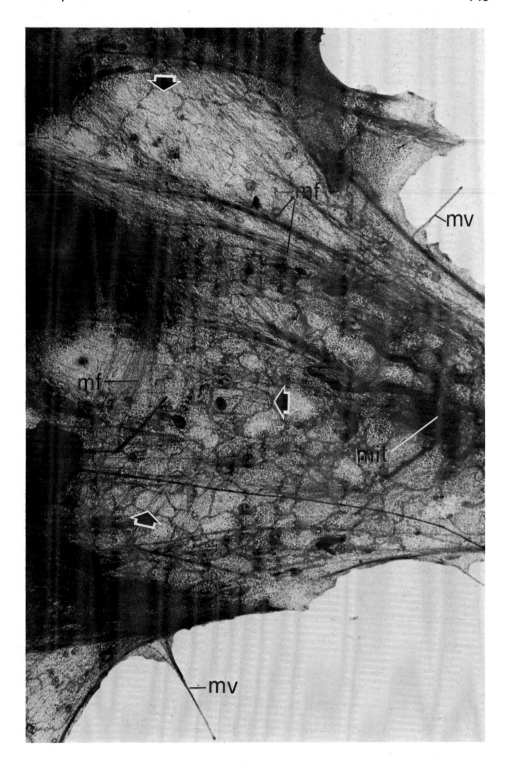

Figure 98

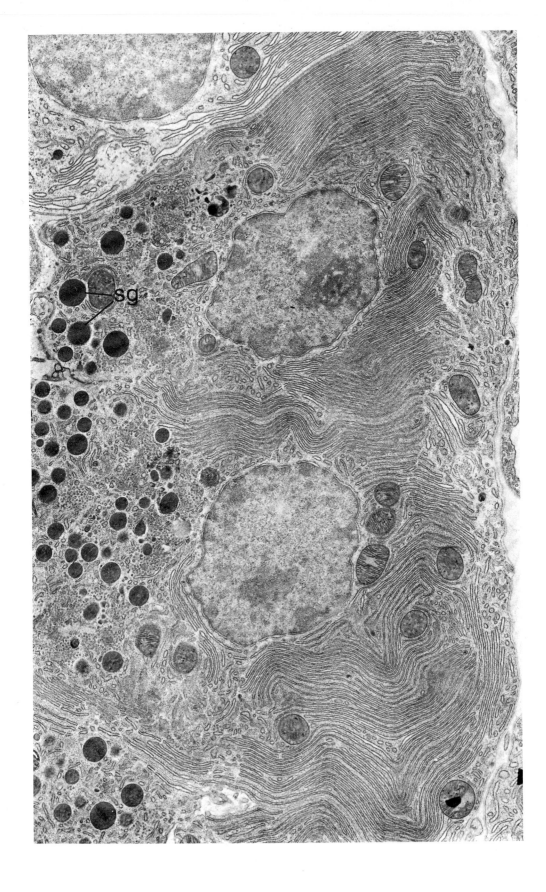

Figure 99). Some idea of the enormously extended surface provided for polyribosome attachment by this arrangement is supplied by quantitative studies of the intracellular morphology of the pancreatic acinar cells. These studies indicate that, although the cisternae occupy only about 20 per cent of the cytoplasmic space, they provide about 65 per cent of the total intracellular membrane surface area.

Functions

Although, as discussed above, the sequestration of newly synthesized proteins is an important function of the rough endoplasmic reticulum, it is clear that, within the cisternae, further processing of the polypeptide translation products can also occur. These alterations may include the formation of disulphide bridges so that the linear polypeptide chain becomes folded and assumes a tertiary (three-dimensional) form, or they may involve glycosylation – the addition of sugar residues like glucosamine and mannose – by specific cisternal transferase enzymes.

Processing at this time may also include degradative steps that reduce the size of the primary translation product. This process is of interest in the regulation of gene expression, since it demonstrates that the modulation and processing of genetic information may still continue even at this stage. At the present time, the best evidence for regulation relates to the processing of peptide hormones, since several of these   *Hormones* (e.g. parathyroid hormone, adrenocorticotrophin, gastrin) are now known to arise initially as part of a much larger, prohormone molecule. In some instances it is even possible that one large primary translation product may be degraded to yield more than one kind of active hormone molecule. The reasons for this arrangement are not clear, but it does seem as if the polyribosomal translation mechanism is only able to deal with mRNAs above a certain size. Thus, if small peptide products (i.e. less than 50 amino acids) are to arise by a process of translation, perhaps they have always to emerge from the polyribosome as part of a larger, precursor molecule.

For studying the transport of luminal contents out of the rough endoplasmic reticulum, autoradiography and subcellular fractionation techniques have both been used in several systems to great advantage. In these studies, a pulse/chase protocol (see page 47) has been employed to follow the introduction of radioactively labelled secretory proteins into the rough endoplasmic reticulum so that their rate of appearance and their clearance can then be followed. In most secretory cells there seems

---

Figure 99. *(Opposite.) Acinar cells in the exocrine pancreas. These cells synthesize and store 'zymogens' – the protein and glycoprotein precursors of pancreatic juice enzymes. Following a meal, the acinar cells secrete their store of granules (sg) which contain the zymogens, and are then able to replenish them within hours. This impressive capacity for protein synthesis depends upon the elaborate parallel array of rough endoplasmic reticulum distributed throughout the basal regions of the cytoplasm.*

Courtesy of K. R. Porter and M. Bonneville, Department of Molecular, Cellular and Developmental Biology, University of Colorado.

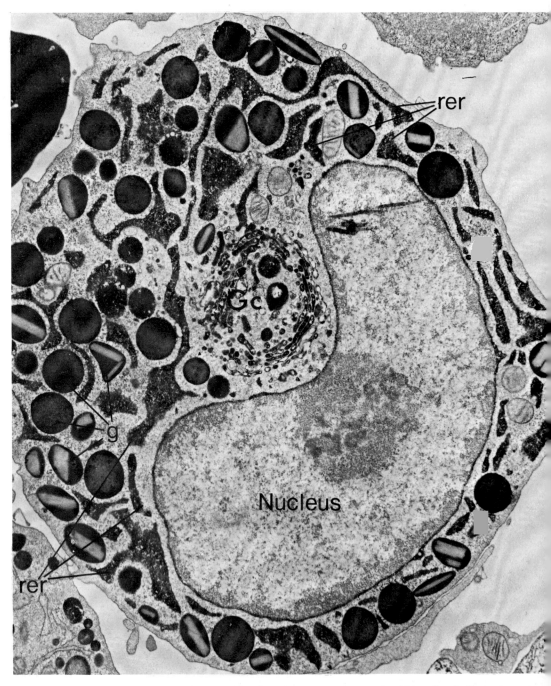

Figure 100. *An electron micrograph of a developing eosinophil leucocyte, stained histochemically to show the intracellular distribution of myeloperoxidase. The peroxidase is synthesized on the polyribosomes of the rough endoplasmic reticulum (rer), and is then transported, via the Golgi complex (Gc), to the spherical cytoplasmic granules (g). It is clear that, at this stage in the development of the cell, all of the cisternae of the rough endoplasmic reticulum (including the nuclear envelope) contain peroxidase. The significance of the crystal-like inclusions within the cytoplasmic granules is not known. Magnification ×8500.*
    Courtesy of D. F. Bainton, Department of Pathology, University of California.

to be labelled secretory product in the rough endoplasmic reticulum within about five minutes of the radioactive amino acid precursor appearing at the cell surface. It is then efficiently cleared to the Golgi complex within about thirty minutes (see Figure 113).

In some kinds of cell it seems that most of the cisternae of the rough endoplasmic reticulum function as a single, homogeneous cell compartment. This situation is best exemplified by the acinar cells of the exocrine pancreas. There, although the rough endoplasmic reticulum is concerned with producing a dozen or more widely different kinds of secretion (which together make up the digestive enzyme constituents of pancreatic juice), all of these products of the attached polyribosomes probably follow the same route out of the cisternae (via the Golgi complex — see below) because all of them have the same intracellular destination (that is, they are packaged, as a mixture, into the same secretory granules).

In many cell types the situation is more complex; as, for example, in the developing eosinophil leucocyte, where two kinds of secretory granule, each with its own distinctive contents, are produced. In this instance, however, the two batches of secretory product are processed at different times in the development of the cell, so that the rough endoplasmic reticulum is primarily concerned with only one group of products in each phase. One of the proteins packaged by eosinophils during the later phases of processing is myeloperoxidase, and this enzyme, like other peroxidases, can be identified histochemically. As shown in Figure 100, at the time when myeloperoxidase is being synthesized in the eosinophil, it is distributed throughout the entire cisternal system.

These arrangements for handling diverse groups of newly synthesized proteins within the endoplasmic reticulum do not, however, account satisfactorily for the requirements of cells that handle products with several different destinations simultaneously. In the liver parenchyma cell, for example, secretory products (i.e. serum proteins), lysosomal enzymes, peroxisomal enzymes, plasma membrane components, and constituents destined for the smooth endoplasmic reticulum all seem to be processed at the same time. Some of these components may be segregated within the cisternae and lie either within the lumen or on the membrane. Nevertheless, in such complex circumstances it seems that, in addition to this kind of partitioning, separate subpopulations of specialized cisternae must also exist.

Autoradiography and cell fractionation indicate that the removal of newly synthesized protein from the rough endoplasmic reticulum is a relatively rapid and efficient process, but as yet the mechanisms responsible have not been identified. It may be that some proteins are cleared by 'membrane flow' (discussed below), but the other obvious alternative, that they are displaced by the products of continuing synthesis moving them on from behind, seems unlikely, since their rate of clearance is only marginally affected when further synthesis is inhibited.

In secretory cells, such as those of the exocrine pancreas, the transport of content out of the rough endoplasmic reticulum almost certainly occurs by 'vesicular evagination'. This is a process of membrane fusion and fission that produces small (40 nm) vesicles in a manner, in molecular terms, that is probably very similar to that which occurs at the plasma membrane during cytosis (see page 85). Vesicular evagination seems to require a continuous supply of energy, since it is readily inhibited by compounds that reduce the availability of ATP.

Membrane biogenesis

The constituents of the plasma membrane and the intracellular membrane systems have limited lives and they are constantly being replaced. For example, although the lifetime of a liver parenchyma cell is between six and twelve months, that of the proteins in its rough endoplasmic reticulum is somewhere between one and twenty hours. The majority of the membrane proteins of the cell are probably products of the polyribosomes attached to the cisternae of the rough endoplasmic reticulum, and many of the lipid components may also be synthesized in this compartment.*

The precise mode of transport of membrane components from the rough endoplasmic reticulum to their destination in membranes elsewhere in the cell remains to be established, but one working hypothesis, for which there is good experimental support, suggests that many of them are transported by membrane flow. Thus, instead of the discrete components being independently synthesized, transported and inserted as replacements in distant and separate membranes, it is likely that new membrane is synthesized initially as part of the cisternal membrane of the endoplasmic reticulum. It is then transported to other membrane boundaries by 'flowing', as part of the membrane lamella (Figure 101).

It follows from this hypothesis that the membranes of a cisternal compartment like the Golgi complex (see Figure 112) will represent a form of mosaic, made up of membranes received from the rough endoplasmic reticulum which are en route for other membrane boundaries such as the cell surface. The other constituents in the mosaic will, of course, be those that allow the particular cisternal compartment to fulfil its own, individual functions.

Evidence in favour of the idea that a membrane flow occurs from the rough endoplasmic reticulum through to the plasma membrane is that the observed asymmetric distribution of the components in these different membranes conforms with that expected of them. Thus, in the cisternal membranes in the rough endoplasmic reticulum of the liver parenchyma cell, the sugar-bearing, N-terminal ends of the glycoproteins are internal. In the Golgi complex, the insulin receptors have a similar orientation. As Figure 101 indicates, this asymmetric arrangement

---

* It is perhaps worth emphasizing that, although any discussion of the rough endoplasmic reticulum necessarily concentrates on the processes of translation and segregation, the cisternal membrane of the reticulum represents a complex organelle in its own right. Some indication of its diverse responsibilities is given by the fact that it displays as many as 40 different enzyme activities.

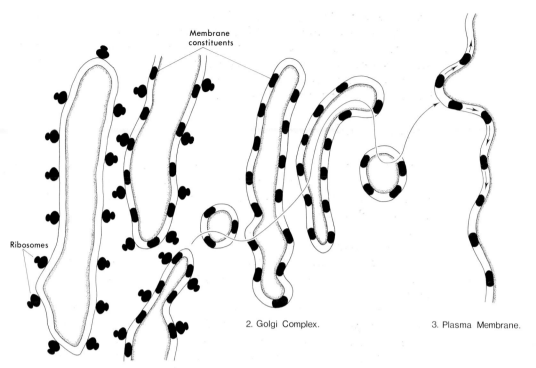

Membrane constituents

Ribosomes

2. Golgi Complex.

3. Plasma Membrane.

1. Rough Endoplasmic Reticulum.

Figure 101. *An outline of the pathway followed by newly synthesized plasma membrane constituents according to the 'membrane flow' model. Components on the inner surface of the intracellular membranes become distributed on the extracellular surface. Note that the membrane thickness is considerably exaggerated.*

is essential if these components are eventually to become (as they are known to be) distributed on the extracellular surface of the plasma membrane.

The mechanism responsible for inserting growing polypeptide chains from the polyribosome across the cisternal membrane of the endoplasmic reticulum might also be of significance in this context, as it may provide a convenient step for the insertion of the amphipathic proteins that span the lipid bilayer. For these components to achieve their transmembrane arrangement, it would only be necessary for their translation to be terminated when their insertion was partially completed and their sugar-bearing termini were penetrating into the cisternal lumen. At the present time these suggestions are clearly very speculative and in need of firm experimental support.

THE AGRANULAR OR SMOOTH-SURFACED ENDOPLASMIC RETICULUM

In many cell types there is a major cisternal component that is closely related to the rough endoplasmic reticulum but has functions other than protein synthesis. This component does not, therefore, bear polyribosomes, and because of this it is given the general heading of 'smooth' endoplasmic reticulum. It is clear, however, that while this is a term of convenience to the morphologist, it represents a great over-simplification in functional terms. In different cell types smooth-surfaced cisternae may perform very different functions.

As a rule, the cisternae of the smooth endoplasmic reticulum are tubular rather than sac-like, and, although their overall contour is usually smooth, they frequently branch. In many situations these cisternae are directly confluent with the cisternae of the rough endoplasmic reticulum, and there is good biochemical evidence to show that, in their biogenesis, the smooth-surfaced cisternae depend for their protein and probably lipid components upon the rough (polyribosome-bearing) cisternae. It is unlikely, however, that there is free intermixing of content between the two systems.

Direct continuity between the smooth endoplasmic reticulum and the Golgi complex (see Figure 112) is less likely, although, in cells such as those of the intestinal mucosa, it is probable that both membrane and content of the smooth-surfaced cisternae can be transferred to the Golgi complex in the form of small (about 40 nm diameter) 'shuttle vesicles' (see Figure 112).

The smooth endoplasmic reticulum and lipid synthesis

In many cell types a major responsibility of the smooth endoplasmic reticulum is the synthesis and transport of lipid. This, however, is not an exclusive responsibility of this cisternal element because the rough endoplasmic reticulum is also known to participate in lipid synthesis.

Steroids

Cells synthesizing steroids

The most extensive development of smooth endoplasmic reticulum concerned with lipid metabolism is found in cells which synthesize steroids. These cells include those of the adrenal cortex, which synthesize a variety of steroid hormones, and those of the reproductive systems (such as the ovarian corpus luteum), which synthesize sex steroids (e.g. progesterone). However, although the biosynthetic pathways for steroids are well characterized, it is not as yet possible to relate their distribution to specific domains within the smooth-surfaced cisternae.

Triglyceride synthesis in cells lining the small intestine

A diet rich in lipid induces extensive development of the smooth endoplasmic reticulum in the apical cytoplasm of the columnar epithelial cells that line the small intestine. In these cells this cisternal component is

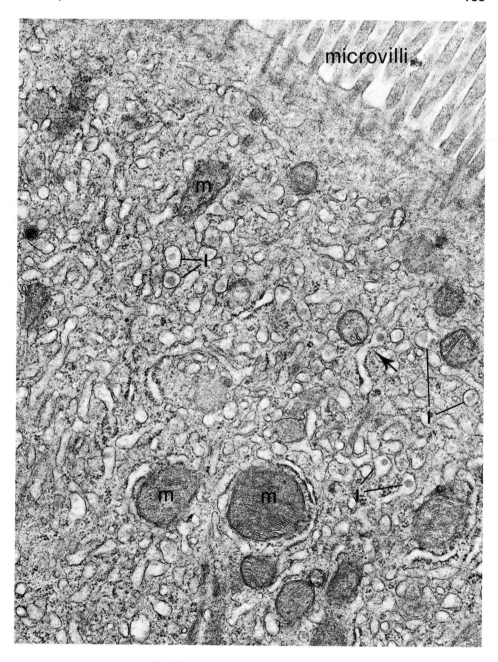

Figure 102. *The apical cytoplasm in a cell lining the small intestine; note the microvilli of the brush border in the upper right-hand corner. In the cytoplasm below the 'terminal web' (see page 226) there is an extensive development of smooth-surfaced, tubular cisternae. Within these elements, triglycerides are synthesized from the small-molecule precursors that are absorbed across the plasma membrane of the microvilli. The newly synthesized lipids associate with proteins derived from the rough endoplasmic reticulum, to form dense lipoprotein droplets (l).*

A profile showing that the rough- and smooth-surfaced cisternae are indeed confluent is indicated by the arrow. m — mitochondria. Magnification ×30 000.

responsible for re-synthesizing lipid from the products of fat digestion absorbed across the plasma membrane of the microvillous border from the lumen. The newly synthesized lipid is sequestered within the lumina of the smooth-surfaced cisternae, and there it can be identified by electron microscopy as discrete spherical particles (Figure 102). Within these cisternae the lipid becomes complexed with a protein component that is presumably derived from the rough-surfaced cisternae.

Chylomicron

The lipoprotein particles that appear within the endoplasmic reticulum are next transported to the Golgi complex, where their transformation into so-called 'chylomicrons' is completed. As indicated in Figure 112, the Golgi complex finally packages the chylomicrons into membrane-bound vesicles and transports them to the lateral cell border, en route for the bloodstream.

The smooth endoplasmic reticulum and detoxification

The cells of the liver parenchyma contain an extensive smooth endoplasmic reticulum, and it is clear that this component has a major responsibility for lipid synthesis. However, in addition, it also plays an important role in drug detoxification, by participating directly in either the breakdown or the chemical modification of toxic substances absorbed across the plasma membrane from the bloodstream. These substances may include toxic chemicals (e.g. herbicides and pesticides), carcinogens, and therapeutic agents (such as barbiturates).

In response to the absorption of a barbiturate by the liver parenchyma cell, there is induced, in the rough endoplasmic reticulum, an increased synthesis of the oxidative enzymes capable of degrading the barbiturate. These enzymes are then transferred to the smooth-surfaced cisternae, where the oxidation occurs (see Figure 103). Initially in barbiturate detoxification there is thus a rapid proliferation of the rough endoplasmic reticulum, which can be correlated with the increased production of the required oxidative enzymes, followed by a proliferation of the smooth endoplasmic reticulum, the site of detoxification. Unfortunately, the result of detoxification is not always beneficial, because some of the degradation products may themselves cause extensive liver cell damage.

Other functions of the smooth endoplasmic reticulum in the liver cell are less clearly understood. For example, although there is consistently

Figure 103. (*Opposite.*) *A low-magnification electron micrograph to illustrate the cytoplasmic organization in a liver parenchyma cell. Throughout the cytoplasm there is an interconnecting cisternal network which includes both rough and smooth elements. The arrows indicate points at which the two kinds of cisterna are confluent. In addition, there are large numbers of mitochondria, several peroxisomes (P) (see page 187), and a few lipid droplets (L). The matrix also contained large amounts of glycogen, but during preparation this has dissolved away. Magnification ×12 000.*

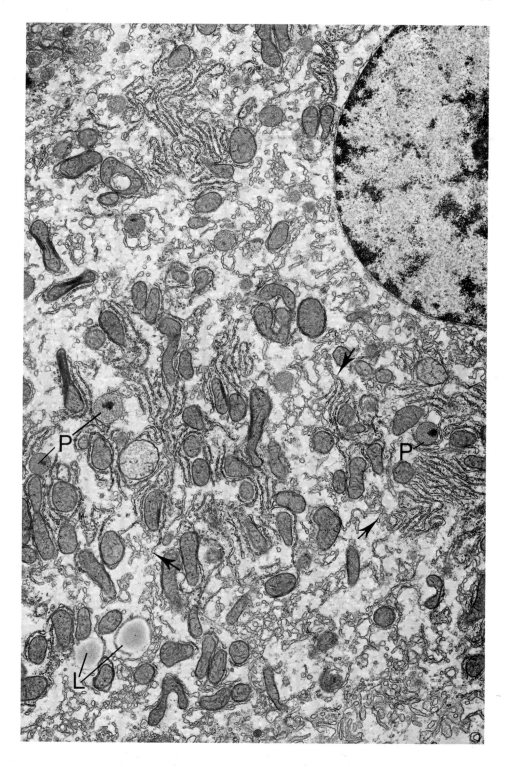

Figure 103

a close association between the distribution of this organelle and gly-cogen storage particles, a clear functional relationship between these two components remains to be established.

The smooth endoplasmic reticulum and ion transport

Calcium as an intracellular signal

As described on page 78, the generation of an action potential across the plasma membrane depends upon an increase in membrane perm-eability that allows sodium ions to flow into the cell down the concentra-tion gradient. A similar change in permeability in respect of the divalent cations may also occur. For calcium ions, in particular, a change in the permeability of the plasma membrane leads to a rapid influx of this cation into the cell, because, although its external concentration is of the same order of magnitude as that of other major cations (i.e. in the millimolar range), its free concentration in the cytoplasm is about a thousand times lower (0.1 to 0.2 micromolar). In most cells this low cytoplasmic level is believed to be due primarily to the calcium-concentrating activity of mitochondria (see page 202).

The influx of calcium ions is probably an important and widely used intracellular signal. It is, for example, known to be important in initiating muscle contraction, and there is good reason to believe that it acts as a 'second messenger' for the release of many kinds of secretion. There is also increasing evidence that calcium provides the trigger signal for initiating DNA synthesis prior to cell division. In many systems the source of the calcium is extracellular, but this is not invariably so. In some cells, for example, the mitochondria probably represent an important alterna-tive, internal source, whereas in striated muscle fibres there is an exten-sive development of smooth endoplasmic reticulum (usually called the 'sarcoplasmic reticulum') that is specialized for this purpose.

The sarcoplasmic reticulum

Skeletal muscle
Smooth muscle

This smooth-surfaced cisternal element is most elaborately developed in the striated muscle fibres of skeletal muscle. In cardiac muscle it is a major but much less well-developed component, whereas in smooth muscle cells it is probably represented by a minor population of short cisternae which is distributed immediately below the plasma membrane (Figure 152). In muscle fibres other than those of skeletal muscle, the transport systems of the plasma membrane play a major role in regu-lating the intracellular calcium level.

In skeletal, striated muscle fibres the sarcoplasmic reticulum is distri-buted around each bundle of 'myofilaments' (i.e. around each 'myo-fibril') in a manner analogous to a fenestrated insulating jacket around a hot-water cylinder (Figures 104 and 105). The cisternae run in the

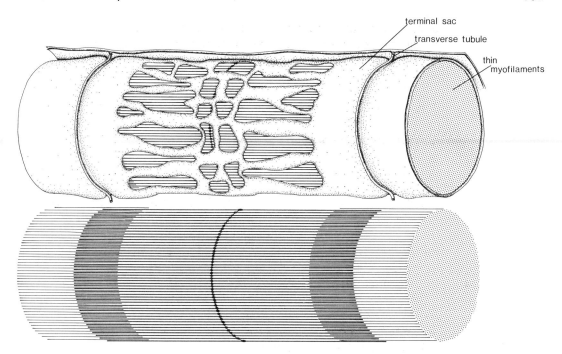

terminal sac

transverse tubule

thin myofilaments

Figure 104. *A diagram showing a portion of the sarcoplasmic reticulum in a striated muscle fibre. In this region of the myofibril the terminal sacs of the sarcoplasmic reticulum surround the area of overlap between the thick and thin filaments. Each sac is closely associated with a transverse tubule which runs on into each fibre across other terminal sacs. For further details on the arrangement of the myofilaments, refer to Figures 141 to 145. Note, there are two transverse tubules to each sarcomere.*

long axis as a network of occasionally branching tubules and, at intervals, in the transverse axis as larger, more sac-like terminal elements. These sac-like elements are confluent with the longitudinal tubules. The terminal cisternae are regularly arranged along the 'sarcomere', so that there are two at each junction of the thick and thin filaments of the myofibril (see page 209). These cisternae are closely associated (but are not in open continuity) with regular inflections of the plasma membrane known as 'transverse tubules'. These tubules reach deep into the muscle fibre.

The primary role of the cisternae of the sarcoplasmic reticulum is to regulate the calcium level in the environment surrounding the myofibrils. The cisternae are able to store the calcium in a bound form. When a muscle fibre is stimulated (see page 78), the depolarization and subsequent increase in Na^+ permeability in the plasma membrane in some way increases the permeability of the sarcoplasmic reticulum (especially in the terminal, sac-like elements). The store of Ca^{++} is then released in free, ionic form into the cytoplasmic matrix, where it induces the interaction between the contractile proteins of the myofilaments that leads to contraction. The distribution of the transverse tubules of the plasma membrane and the terminal cisternae of the sarcoplasmic reticulum that surround the thick and thin myofilaments in the all-important region of

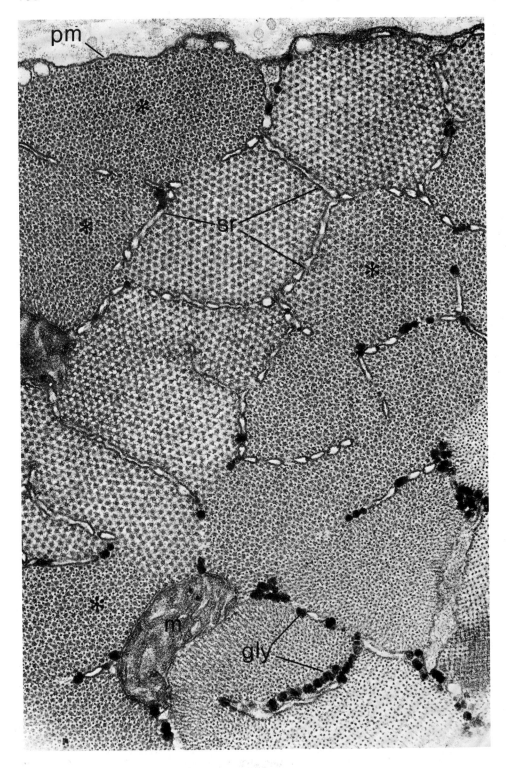

Figure 105

their overlap (see Figure 104) ensures that the initiating events occur virtually simultaneously throughout the muscle fibre. Once stimulation ceases, the membranes of the sarcoplasmic reticulum actively transport the calcium back into the cisternal lumen, and relaxation ensues.

Structurally, the membranes of the sarcoplasmic reticulum are relatively simple, since they contain only two integral proteins. These proteins are a calcium-activated adenosine triphosphatase (ATPase), which constitutes about 95 per cent of the total integral protein mass, and a proteolipid, which constitutes about 5 per cent. The ATPase (mol. wt. about 100 000) is amphipathic (see page 63); it penetrates the central lipid bilayer and appears in freeze-fractured replicas as an 8 nm globule facing the cytoplasmic surface. Using the energy derived from splitting ATP, it is able to transport calcium across the membrane and into the cisterna (one ATP is required for every two Ca^{++} ions transported).

On the inner surface of the cisternal membrane are two peripheral glycoproteins (mol. wt. about 55 000 — one called 'calsequestrin') that are able to bind calcium ions. These proteins allow the cisternae to store calcium at high concentration, equal to a free ionic concentration of between ten and twenty millimolar.

Recent studies suggest that a simple membrane organization, consisting only of a lipid bilayer and a two-protein complex capable of transporting and sequestering calcium, may be present in cells other than those of muscle. For example, similar proteins with the same molecular weight and the ability to concentrate Ca^{++} ions have been identified in the clathrin-coated vesicles (see page 185) of nervous tissue.

The smooth endoplasmic reticulum and fragmentation of the cell periphery

Amongst the blood-borne, cellular elements produced in bone marrow are the platelets. These cellular fragments play an important role in blood clotting and, although they lack nuclei, they contain mitochondria and a variety of secretory granules. In bone marrow they arise by continuous fragmentation, from the periphery, of large, polyploid cells known as megakaryocytes and this process relies upon an extensive development of smooth-surfaced cisternae throughout the megakaryocyte cytoplasm (Figures 106 and 107). The cisternae probably serve as demarcation channels and, like the perforations in a page of postage stamps, they indicate the line of fragmentation which will give rise to the platelets. Probably the connection between the forming platelet and the parent

---

Figure 105. (*Opposite.*) *A cross section of a striated muscle fibre to show the distribution of the sarcoplasmic reticulum (sr). The section plane transects the myofibrils close to the junction between the A- and I-bands, where the thick and thin myofilaments overlap (see Figure 143 and page 209). Thick and thin filaments are seen in myofibrils marked by an asterisk. pm — plasma membrane; gly — glycogen particles; m — mitochondria. Magnification × 62 500.*
Courtesy of C. Peracchia, Department of Physiology, University of Rochester.

Figure 106. *A low-magnification view of the cytoplasmic organization in the perinuclear region of a megakaryocyte. The cytoplasm contains an elaborate network of smooth endoplasmic reticulum, which divides the cell into small, platelet-sized portions. This cell type is by far the largest in the bone marrow. During its development, the nucleus divides and then re-fuses several times to form a large, multilobed structure. Several lobes appear in this section plane. Magnification ×4000.*

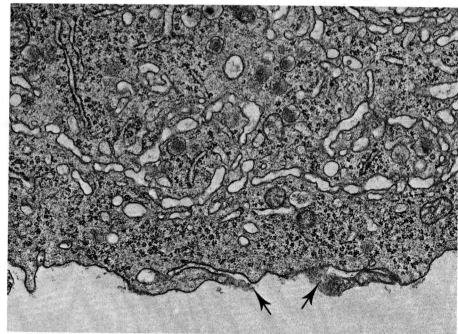

Figure 107. *At the periphery of this megakaryocyte there are clear indications (arrows) that fragmentation occurs as a result of the smooth-surfaced cisternal membrane fusing with the plasma membrane. Magnification ×24 000.*

megakaryocyte is severed by a process of membrane fusion and fission similar to that which occurs during cytosis. The integrity of the mega-karyocyte boundary is thus maintained.

THE VACUOLAR SYSTEM

Loosely collected under this heading are the cisternal elements of the Golgi complex and the vesicular and vacuolar elements (like lysosomes and secretory granules) to which they give rise or are otherwise associated.

THE GOLGI COMPLEX

Until the details of the form and distribution of the Golgi complex were observed in the electron microscope, direct evidence for even its existence depended primarily upon the capricious and little-understood

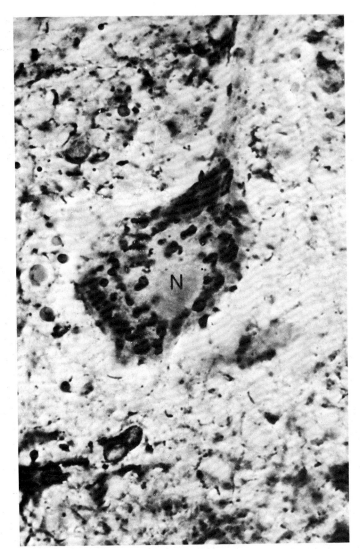

Figure 108. *The classical Golgi silver impregnation technique. Within the ventral horn neurone shown here, the Golgi complex surrounds the nucleus (N) as a densely impregnated, tortuous reticulum. Magnification ×800.*

staining methods devised by Camillo Golgi (1843–1926), its discoverer (Figure 108). Now, although the almost ubiquitous distribution of this organelle has been clearly established, it still remains one of the more enigmatic features of the cell. This is because of its complex structure and diverse functional activities, and the difficulty of studying it in isolation.

In cells with a polarized intracellular organization, such as the columnar epithelial cells of the intestinal mucosa or the acinar cells of

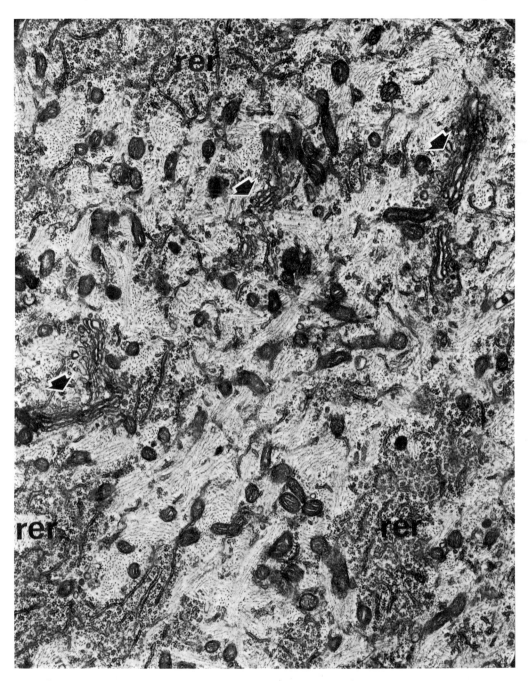

Figure 109. *The organization of the cytoplasmic organelles in the cell body of a neurone. The patchy distribution of the rough endoplasmic reticulum (rer) is very characteristic, and readily related to the Nissl bodies seen in conventional light microscope preparations (Figure 87). Golgi complex components are distributed across the centre of the field (arrows) and they are presumably part of an elaborate system similar to that outlined in Figure 110. In the cytoplasmic matrix surrounding the abundant mitochondria there are loose bundles of neurofilaments (see page 233). It should be noted that neurofilaments are not such a prominent feature in all neurone cell bodies. Magnification ×20 000.*

Courtesy of S. L. Palay, Department of Anatomy, Harvard Medical School.

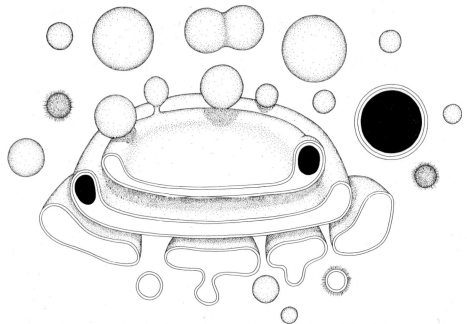

Figure 110. *A three-dimensional reconstruction of a stack of flattened Golgi cisternae to show the saucer-like form of the cisternal elements. Shuttle vesicles, derived from the rough endoplasmic reticulum, fuse and contribute their unconcentrated content to the outer, most voluminous cisterna. Secretory granules usually bud off at the periphery of the inner (releasing face) cisterna.*

the exocrine pancreas, the Golgi complex lies within the apical cytoplasm immediately adjacent to the nucleus. Often it surrounds that part of the cytoplasm which contains the 'centrioles', called the 'cytocentrum' (see page 238). However, no special functional relationship between the Golgi complex and this region of the cytoplasm has yet been demonstrated.

In electron micrographs, the Golgi complex is seen to consist of an extremely variable population of flattened, sometimes tubular cisternae and their associated membrane-bound vesicles and vacuoles. It is not clear if any of the cisternal elements are in direct continuity with those of the endoplasmic reticulum, although it has been established (see below) that vesicular elements arising from the rough endoplasmic reticulum can fuse and contribute their content to the Golgi cisternae. It is very probable that the Golgi complex includes several distinct cisternal populations, and in some cell types (for example, in nerve cell bodies – see Figure 109) they form an extensive, interconnecting network throughout the cytoplasm.

Cisternae. Closed, membrane-limited sacs.

Despite its complexity, however, there is a consistent cisternal arrangement amongst the Golgi elements of most cells. This arrangement consists of a stack of three to twelve flattened, saucer-like cisternae surrounded by a variable number of spherical 60 to 80 nm vesicles (Figure 110). Thin sections cut across this cisternal stack indicate that its cisternae arise at the convex or 'forming' face and move, as they become

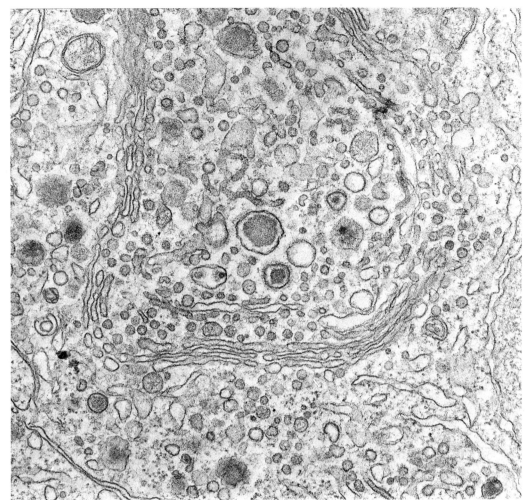

Figure 111. *The Golgi area in a pituitary secretory cell. The section plane transects the releasing face of the complex and illustrates that, in addition to the characteristic and rather consistent features shown in Figure 110, there may be a wide range of other cisternal and vesicular components within the area. Magnification ×30 000.*

displaced by their successors, towards the concave or 'releasing' face. At the releasing face the flattened cisternae break up to give rise to a variety of membrane-bound Golgi derivatives, such as secretory granules and, probably, primary lysosomes.

The flattened cisterna at the convex, forming face of a Golgi cisternal stack usually has a voluminous, electron-lucent content and probably arises by the fusion and coalescence of vesicles largely derived from the rough endoplasmic reticulum. In this context, it is interesting to note that the thickness of the cisternal membrane at the forming face is similar to that of the endoplasmic reticulum (about 5 nm), while that of the releasing face is thicker and more like that of the plasma membrane (about 7.5 nm). The cisternae at and near the concave, releasing face of the complex are usually flattened and attenuated, except at the rim, where they may be variously inflated to form bulb-like protrusions. In the vicinity of the most concave cisternae there is usually a heterogeneous population of membrane-bound vacuoles and, depending on the cell type, secretory granules (Figure 111). These structures are derived from the bulbous protrusions of previous cisternae.

In many cell types the forming face of the Golgi stack is directed away from the nucleus towards the cell surface. The polarity of the stack, however, is thought to have a functional rather than a purely structural basis, because in some cells the direction of transport may change. In the developing neutrophil leucocyte, for example, two kinds of granule arise from the Golgi complex; the first-formed (the developing 'azurophil' granules) arise from the face nearest the nucleus, while the later-formed (the specific or 'neutrophil' granules) are derived from the opposite face.

Functions

1. Membrane redistribution

The interrelationships that exist between the Golgi complex, rough endoplasmic reticulum, plasma membrane, and membrane-bound derivatives of the Golgi complex (such as secretory granules) indicate that the Golgi complex is a major centre for the redistribution of cell membranes. As shown in Figure 112, the complex receives membrane-bound vesicles from the rough endoplasmic reticulum at its forming face, and provides membrane to the plasma membrane via the enveloping membranes of the vesicular elements that arise at the releasing face. Similarly, membrane-bound vacuoles formed at the plasma membrane by endocytosis follow the reverse route, from the cell surface into the Golgi area.

Membrane routes such as those between the Golgi complex and the plasma membrane may well provide a round trip for membranes (and to some extent content), but this is not true of all routes. In particular, it seems that the pathway between the rough endoplasmic reticulum and the Golgi complex is uni-directional, since membrane and cisternal content pass only towards the Golgi complex. Nothing entering the Golgi complex or any other cisternal compartment (with the exception of the nuclear membrane, which is a special case) ever seems to gain access to the cisternae of the rough endoplasmic reticulum. The specificity that determines which membrane component will fuse with one but not another neighbouring component remains to be identified.

2. Concentration of secretory products

Exocrine

Endocrine

In some secretory cell types (most notably those of the exocrine pancreas and parotid gland and the endocrine cells of the pituitary and adrenal medulla), secretion granules containing secretory product arise at the

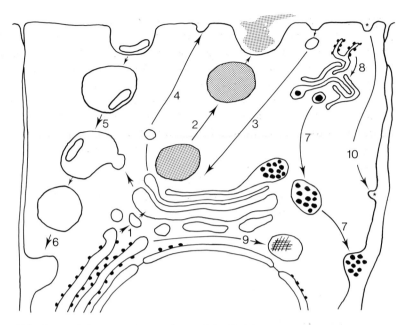

Figure 112. *Some of the routes into and out of the Golgi area:*

1. Cisternal content and membrane, arising on attached polyribosomes, are contributed to the Golgi complex by the rough endoplasmic reticulum.
2. Secretory product is packaged into vesicles by the Golgi complex, and is transported to the plasma membrane for release by exocytosis.
3. Compensatory withdrawal of membrane from the plasma membrane.
4. A contribution of membrane components to the cell surface.
5. Endocytosis, and the formation of secondary lysosomes (see Figure 122).
6. Exocytosis of the lysosomal content (see Figure 122).
7. Triglycerides, synthesized in the smooth endoplasmic reticulum, are packaged and processed by the Golgi complex into chylomicrons; these are then transported to the lateral plasma membrane for release by exocytosis.

Membrane transfer between some compartments does not necessarily require transport via the Golgi complex:

8. Synthesis of smooth endoplasmic reticulum by the rough endoplasmic reticulum.
9. Formation of peroxisomes (see page 187).
10. Cytotic transport across the cell, as in capillary endothelial cells.

releasing face of the Golgi complex. These granules are then stored, often in large numbers, in the cytoplasm until their release by exocytosis is stimulated. In these cells it is a function of the Golgi cisternae to begin concentrating the secretion product before packaging it into the granules (Figures 113 and 114). Evidence that concentration is occurring (presumably by the transport of water and ions across the cisternal membrane and into the surrounding cytoplasm) is seen first within the expanded bulbs of the inner Golgi cisternae, where the secretion appears as an electron-opaque condensate. However, although the process of concentration often begins in the Golgi cisternae, it usually continues after the secretion granule is formed. As a rule it then proceeds, until, in the mature granule, the content becomes a tightly enveloped electron-opaque mass.

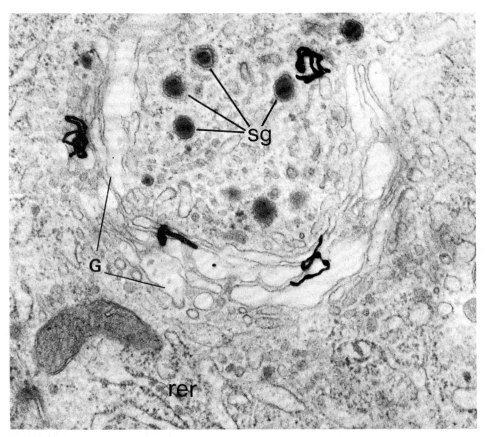

Figure 113. *An electron microscope autoradiograph showing the Golgi area in a secretory cell from the pituitary. The tissue was given a 5-minute pulse of a radioactively labelled amino acid precursor and then incubated for varying intervals post pulse with a high concentration of the same, but unlabelled, precursor. With this pulse/ chase protocol, the secretory product containing the labelled precursor travels from the site of synthesis – in the rough endoplasmic reticulum (rer) – to and through the Golgi complex, where it is packaged into secretory granules (sg). In this autoradiograph, taken at 30 minutes post pulse, the label has reached the flattened Golgi cisternae (G). Its location is indicated by the dense, irregular silver grains. Magnification ×20 000.*

3. Processing of secretory products

Thyroglobulin. This is the molecule within which thyroid hormones are incorporated; it is synthesized and stored in the thyroid gland.

Gammaglobulin. A class of serum glycoproteins; they include immunoglobulins.

The synthesis of glycoprotein molecules destined either for secretion (examples include thyroglobulin and gammaglobulin) or the cell surface begins on the polyribosomes of the rough endoplasmic reticulum. In this compartment, in addition to the amino acid residues that form the polypeptide 'backbone' of the glycoprotein molecules, sugar residues, such as glucosamine and mannose, are also incorporated. These sugars are present within (or close to) the 'backbone' of the glycoprotein. The addition of further sugar residues to the side-branches of the molecular 'backbone' continues as the glycoprotein is transported through the rough endoplasmic reticulum to the Golgi complex. Synthesis is finally completed with the addition of the so-called 'terminal' sugars (fucose

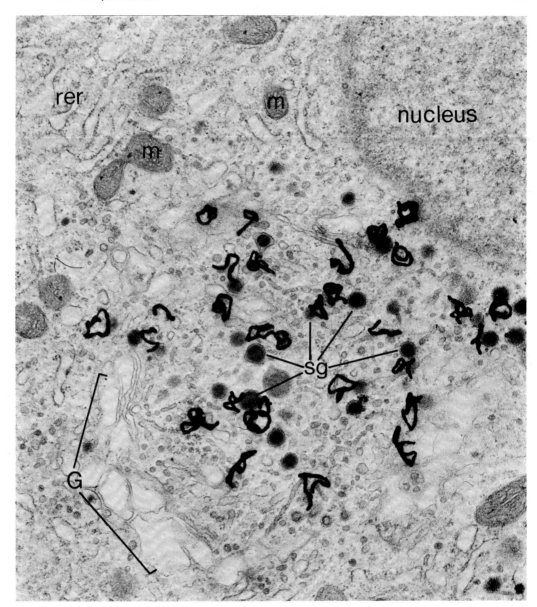

Figure 114. *Preparation as for Figure 113, but this autoradiograph shows the location of the labelled secretory product at 50 minutes post pulse. By this time the label has been chased from the Golgi cisternae (G) into (almost certainly) the secretory granules (sg). Note that although the resolution of the autoradiographic technique is insufficient to identify which vesicular elements of the Golgi complex contain the radioactive label, it is clear that the other cellular components in the area – the nucleus, mitochondria (m), and rough endoplasmic reticulum (rer) – remain unlabelled. Magnification ×15 000.*

and/or sialic acid) within the flattened cisternae of the Golgi. This pathway has been traced in several different cell types by electron microscope autoradiography; parallel, subcellular fractionation studies have shown that the specific transferase enzymes required for the addition of the

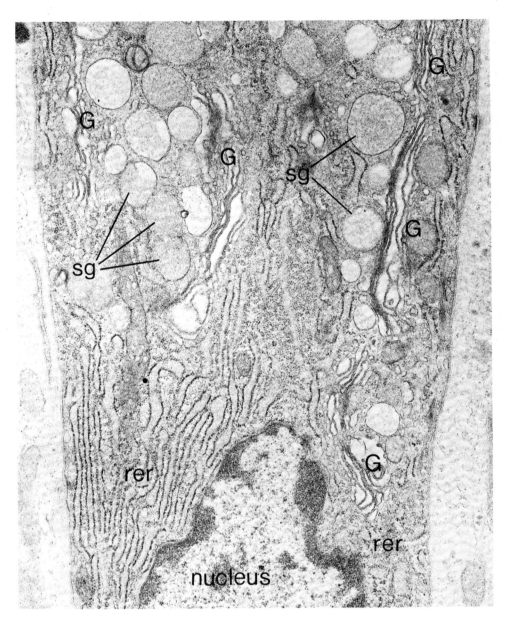

Figure 115. *The Golgi complex of a goblet cell in the small intestine. As in most polarized cells, the Golgi complex is in the apical cytoplasm, immediately above the nucleus. Although most of the rough endoplasmic reticulum lies in the basal cytoplasm, its more apical cisternae (rer) extend around the nucleus and into the Golgi area. The most prominent features of the Golgi complex in this cell are the stacks of flattened cisternae (G). They are almost certainly part of the same cisternal system that folds and refolds in a tortuous array across the width of the cell. Cisternae at the forming face (or faces) are indicated by their voluminous content, while those at the releasing face (or faces) are collapsed and attenuated. The membrane-bound secretory granules (sg) contain mucus, and are seen en route to the apical cytoplasm (where they accumulate ready for release). Magnification ×9500.*

relevant sugars are concentrated in the appropriate compartments. As expected, the Golgi complex is characteristically rich in terminal sugar transferases.

The Golgi complex also participates in the biosynthesis of sulphated mucopolysaccharides. As their name suggests, these substances, which include the mucus synthesized and secreted by goblet cells in the digestive tract (Figure 115) and heparin stored in the secretion granules of mast cells, contain inorganic sulphate. By using the radioisotope $^{35}SO_4$, it is possible to employ autoradiography to demonstrate directly that sulphate addition occurs in the Golgi complex. As might be expected, autoradiography using this isotope indicates that sulphation is a function of the Golgi complex in most cell types, and shows that it is not necessarily restricted to those concerned with the synthesis of copious amounts of mucopolysaccharide.

Another role for the Golgi complex exists in cells in which the latter stages of secretory product processing require limited enzymic (especially proteolytic) cleavage. The best-documented system in this regard is that of the beta cell in the islet of Langerhans, which synthesizes and secretes the hormone, insulin. This hormone is a protein comprising two polypeptide chains linked by disulphide bridges, but when it is synthesized on the polyribosomes of the rough endoplasmic reticulum it is made as a long, single-chain precursor, known as 'proinsulin'. When the precursor (or 'prohormone') leaves the endoplasmic reticulum it is similarly linked with disulphide bonds, and, as shown in Figure 116,

Mast cells.
Common constituents of loose connective tissues.

Insulin

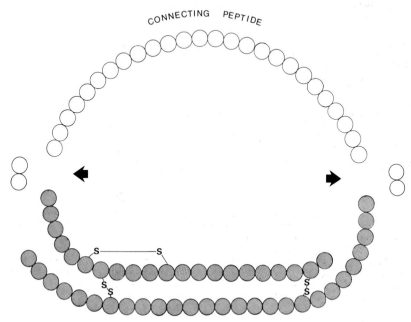

Figure 116. *The severed proinsulin molecule, showing the points at which proteolytic cleavage yields the double polypeptide chain of insulin. After D. F. Steiner.*

the conversion thus requires only an enzymic cleavage at two points to convert the linear, single-chain translation product into the double-stranded (and biologically active) insulin molecule. The cleavage of proinsulin almost certainly takes place in the Golgi complex, and it yields, in addition to insulin, a small peptide fragment known as the connecting or 'C' peptide. This peptide appears to be simply a non-functional by-product of the proinsulin conversion. Both the hormone and the C peptide are then packaged by the Golgi complex into secretory granules.

In some other secretory cells, such as those that secrete the hormone gastrin, enzymic cleavages are required that are of a specificity similar to that shown in the conversion of proinsulin to insulin. It remains to be seen if these cleavages, too, occur in the Golgi complex.

Gastrin. A peptide hormone secreted by a minority population (the 'G' cells) of the gastric mucosa; it provokes gastric acid secretion.

4. Primary lysosome formation

Although it remains to be established beyond all doubt, it is probable that the Golgi complex is also responsible for concentrating and packaging hydrolytic enzymes into primary lysosomes. These structures (as described below) contain a battery of enzymes upon which the cellular

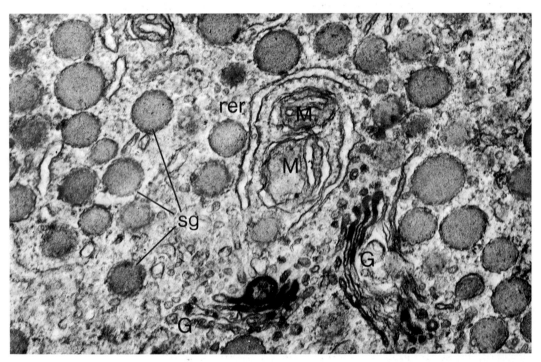

Figure 117. *The edge of a Golgi area in a pituitary secretory cell, stained histochemically for the lysosomal enzyme, acid phosphatase. Although the secretory granules (sg), mitochondria (M), and cisternae of the rough endoplasmic reticulum (rer) are unstained, the flattened cisternae of the Golgi complex (G) contain a dense precipitate of reaction product. Presumably these cisternae produce the primary lysosomes. Magnification ×24 000.*

digestive system depends. Histochemical methods such as the lead phosphate method for acid phosphatase, described on page 38, indicate that at least some of these enzymes are active within the Golgi cisternae (Figure 117). How they remain sequestered from other cisternal contents is not known, although it remains possible that amongst the various subpopulations of cisternae that exist in the Golgi area there are some that are specifically concerned with handling acid hydrolases. Nevertheless, other methods of segregating the different kinds of content within a single cisterna are also probably required, and in this context it is of interest to note that at least some of the lysosomal enzymes are believed to be attached to the cisternal membrane.

LYSOSOMES

This group of interrelated membrane-limited elements forms an intracellular digestive system which is concerned with degrading substances of both extracellular and intracellular origin (Figure 118). Lysosomes are polymorphic, vacuolar structures which contain degradative hydrolytic enzymes that are optimally active at an acid pH. They are invariably membrane-bound. First identified by cell fractionation methods (Figure 119), lysosomes have been shown to contain up to sixty different hydrolytic enzymes, and these, together, are able to degrade a wide range of cellular and non-cellular substances.

Unlike mitochondria and secretory granules, lysosomes have few characteristic morphological features (Figure 120). From the beginning, therefore, studies of their form and distribution have been heavily dependent upon the availability of histochemical techniques, applicable at the electron microscope level, for so-called 'marker' enzymes. The lead phosphate technique for the identification of acid phosphatase, described on page 38, has been the histochemical method most widely used for this purpose (Figure 121).

Heterophagy and autophagy

Substances of extracellular origin enter the vacuolar apparatus following their endocytotic uptake. They thus enter the cell's interior in membrane-bound vacuoles, and since the formation of these vacuoles is the result of 'heterophagy' (cf. 'autophagy') they are termed 'heterophagosomes' (Figure 122). The cell's own components, such as mitochondria, whole secretion granules, or simply small areas of cytoplasm, enter the system by becoming surrounded by smooth-surfaced cisternae, probably derived from the Golgi or the endoplasmic reticulum. This process is termed 'autophagy' and it gives rise to 'autophagosomes' (Figures 122 and 123). Heterophagosomes and autophagosomes only become lysosomal

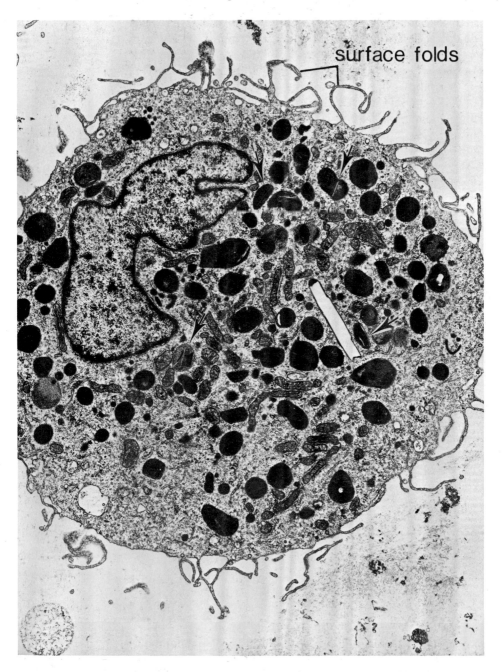

Figure 118. *An electron micrograph of a macrophage from the lung tissue of a cigarette smoker. Amongst the dense population of lysosomal elements there are some containing distinctive, needle-like, electron-lucent inclusions (arrows). These inclusions are found only in the macrophages of cigarette smokers and are believed to consist of 'kaolinite', an aluminium silicate known to be present in tobacco. Magnification ×25 000.*
   Courtesy of A. R. Brody and J. E. Craighead, University of Vermont.

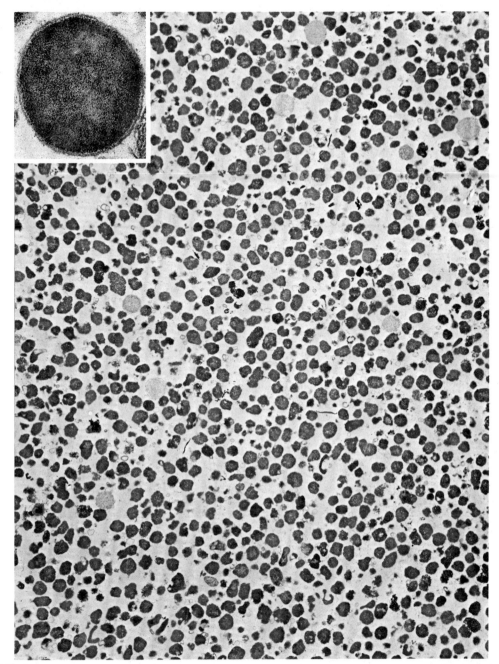

Figure 119. *Lysosomes were first identified as a separate class of organelles in studies using cell fractionation. Although this has continued to be the most widely applied method of preparation, its usefulness has always been compromised by the characteristic heterogeneity of the lysosomal population, which has meant that these organelles share the same size and density range as other subcellular components (especially peroxisomes and mitochondria). For this reason, methods of converting lysosomes into a more easily separated subfraction have been sought. The preparation shown here is a subfraction obtained from the liver of animals fed on an iron-rich diet. In these circumstances, the lysosomes become loaded with an iron-storage product (ferritin) and thus have a greatly increased density. By a combination of differential and density gradient centrifugation, these organelles can now, therefore, be readily separated from other cell components to produce a subfraction that is about 99 per cent lysosomes. Magnification ×5000.*

Shown in the inset is an iron-loaded lysosome with a clearly defined limiting membrane. Magnification ×80 000.

Courtesy of H. Glaumann, Institute of Pathology, Karolinska Institute, Stockholm, and Department of Pathology, Sabbatsberg's Hospital, Stockholm.

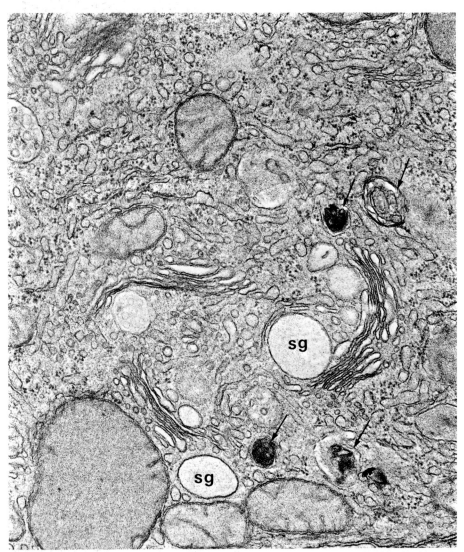

Figure 120. *The Golgi area in an epithelial cell lining the respiratory tract. Within the immediate vicinity of the flattened cisternal stacks there are several kinds of lysosomal element (arrows) and two almost mature secretory granules (sg). In the surrounding cytoplasm there are several mitochondria and an extensive (partly rough) endoplasmic reticulum. Magnification × 40 000.*

when they fuse with lysosomal elements which contain acid hydrolases. If these hydrolase-containing elements have not been previously involved in degradative activity, they are regarded as 'primary' lysosomes. As described above, primary lysosomes probably arise from the Golgi complex, and on fusing and contributing their acid hydrolase content to phagosomes, form 'secondary' lysosomes. It is within the secondary lysosomes that digestion occurs. Contained proteins are degraded to their constituent amino acids, carbohydrates to monosaccharides, nucleic acids to nucleosides and phosphates, and lipids to free fatty acids. These

Figure 121. *Two secondary lysosomes (L) demonstrated by their content of acid phosphatase reaction product. The method of electron microscope histochemistry used for this preparation is described on page 38. In order to preserve the general cellular organization and still retain the activity of the acid phosphatase, a less than optimal concentration of fixative was employed. Nevertheless, it is clear that of the organelles seen here, which include mitochondria (m) and endoplasmic reticulum (rer), only the lysosomes contain acid phosphatase. Magnification ×36 000.*

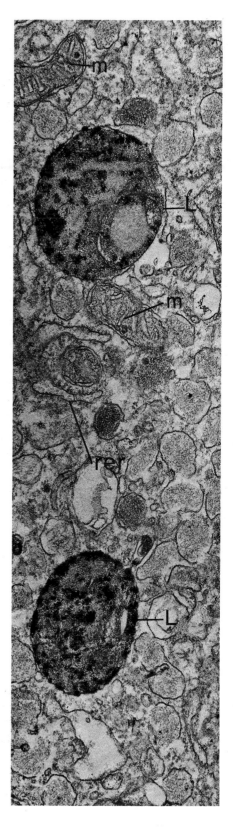

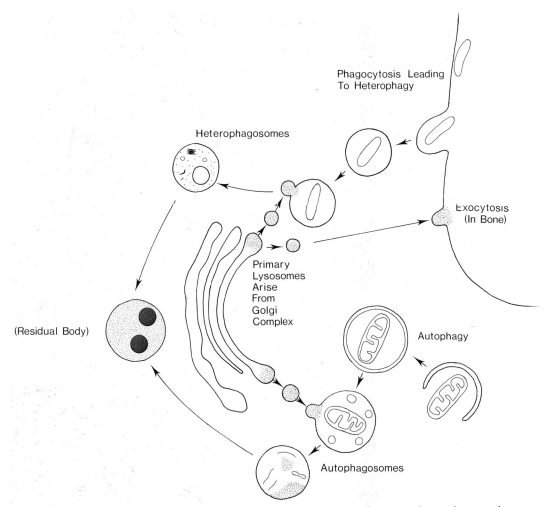

Figure 122. *The intracellular pathways followed by lysosomal elements. For convenience, they are shown as separate routes, but fusions between the different kinds of secondary lysosome are probably very common.*

products are of a molecular size (mol. wt. about 300) small enough to diffuse across the limiting lysosomal membrane into the surrounding cytoplasmic matrix for use by the cell. Following this digestive activity, only undegradable components remain. Most typically they include iron (from haem proteins like haemoglobin) and lipid. The secondary lysosomes, at this stage termed 'residual bodies', may remain and accumulate within the system for extended periods.

Heterophagy is seen most typically in active phagocytes, such as the free tissue macrophages (e.g. in the lung) and the fixed cells of the liver and reticular tissues (e.g. the spleen and lymph node). In these cells the production of primary lysosomes is a continuing process which is probably coupled directly to the cell's phagocytic activity. In the eosinophil granular leucocyte (Figure 124), on the other hand, the large

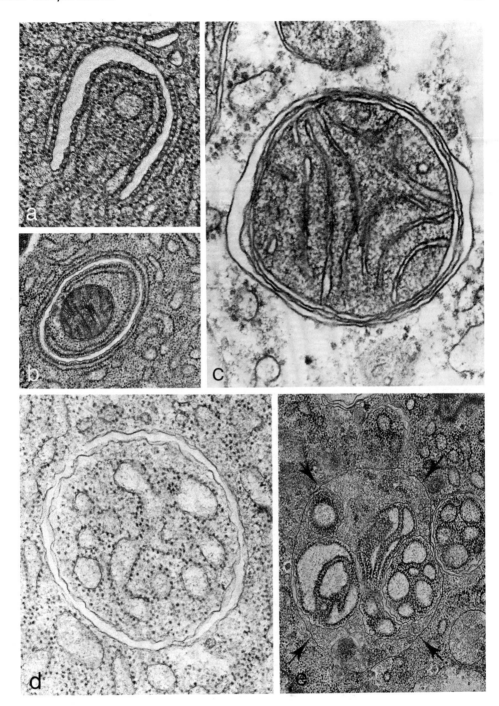

Figure 123. *Profiles indicating the steps involved in autophagy. (a) and (b) Envelopment of cytoplasmic organelles by smooth-surfaced cisternae. (c) and (d) Autophagic vacuoles still bounded by a double-membrane envelope. (e) Larger, more complex structures that have probably arisen by the fusion of smaller autophagosomes (arrows). Magnifications: a ×44 000; b ×26 000; c ×78 000; d ×50 000; e ×25 000.*
   Courtesy of M. Locke, Cell Science Laboratories, Department of Zoology, University of Western Ontario.

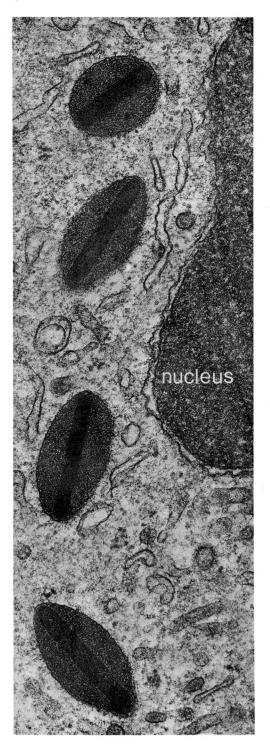

Figure 124. *A small portion of an eosinophil leucocyte, showing the characteristic form of the so-called 'specific' granules. These granules contain acid hydrolases, and therefore represent primary lysosomes. The nature of the characteristic crystalloid structure that lies in their long axis is not known. Magnification ×42 000.*

eosin-positive granules that characterize the cytoplasm of the mature cell represent this cell's full complement of primary lysosomes. When attacking an invading micro-organism, therefore, a granulocyte can immediately draw upon a large lysosomal population. However, these cells have no capacity for replenishing their lysosomal supply, and thus their digestive capacity is clearly limited to a small number of such encounters.

Autophagy occurs in most kinds of cell, but it is found predominantly in the cells of tissues which undergo age-related or hormone-dependent remodelling. Good examples are to be found in the secretory tissues of the reproductive system.

The release of lysosomal enzymes from the cell

Lysosomal enzymes may be released into the extracellular space as a result of exocytosis; they may thus follow the same route from the endoplasmic reticulum to the cell surface, via the Golgi complex, as that taken by other secretory products. The release of lysosomal enzymes in this manner is best documented in the osteoclast, a bone cell concerned with degrading and removing the collagen fibres and inorganic salts of the bone matrix. The osteoclast, when actively resorbing bone, is closely applied to the bone surface and secretes its lysosomal enzymes into the shallow lacuna that develops beneath it. The activity of the enzymes is thus concentrated within this confined area, and the osteoclast is well placed to endocytose and further degrade intracellularly those constituents of the bone matrix that are released.

Lysosomal enzymes are also known to be released into the extracellular space by other actively phagocytic cells, but in these circumstances the released enzymes are not contained in a concentrated form within the immediate vicinity of the cell. It is thus not easy to imagine that they can play a very effective extracellular role, and it seems probable that their release is the result of 'unintentional' leakage, caused by lysosomes fusing with forming heterophagosomes before their contents become completely sealed off from the extracellular space.

The extrusion of lysosomal content does, however, also occur as a means of eliminating undegradable products from the cell. This process occurs in liver parenchyma cells, where, by exocytosis, the so-called 'peribiliary' lysosomes empty their contents into the adjacent bile canaliculus. It is not clear to what extent this process of 'defecation' occurs in other locations. Elsewhere, however, it is unlikely to provide an effective means of waste disposal, for presumably it will only result in re-uptake by neighbouring phagocytes. With time, there is, indeed, an increase in the number of lysosomes containing undigested residues in all tissues. This 'age pigment', as it is often called, can be a particularly prominent feature of nerve and muscle cells.

The role of lysosomes in secretion

In secretory tissues in which the requirement for further secretion becomes reduced (as, for example, occurs in the milk-producing cells of the mammary gland when the suckling young have been weaned), autophagy eventually leads to cell atrophy and loss. Under these circumstances many whole secretory granules, and even groups of granules, will become incorporated into autophagosomes. However, in some active secretory cells where the demand may be reduced abruptly (and the release of secretion inhibited) an interim arrangement may be found which seems to hinder the cell from becoming involved in a large-scale autophagy of its secretory machinery. This arrangement, which is called 'crinophagy', prevents the over-production of secretory granules by funnelling the newly formed granules directly into the lysosomal pathway. In crinophagy, unlike the usual autophagic process, the limiting membranes of the secretion granules fuse directly with those of the lysosomal elements. With a reinstatement of the demand for secretion, mature

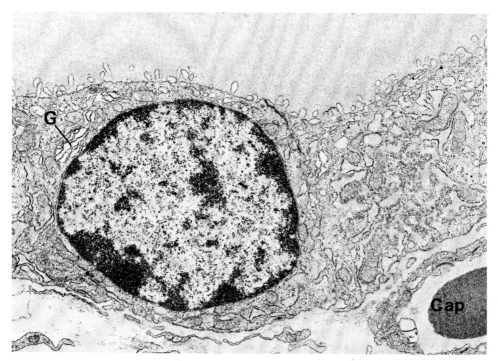

Figure 125. *An electron micrograph showing a portion of a thyroid follicle wall. The follicle cells contain abundant rough endoplasmic reticulum and well-developed Golgi complexes (G). Inside the follicle lumen the thyroglobulin produced by these cells is stored within a homogeneous extracellular matrix, the colloid. Outside the follicle there is a well-developed bed of blood capillaries (Cap).*
   Since this particular follicle is characterized by a large amount of colloid and a narrow follicular epithelium (containing few lysosomal elements), the rate of thyroglobulin breakdown is probably low. Magnification ×10 500.

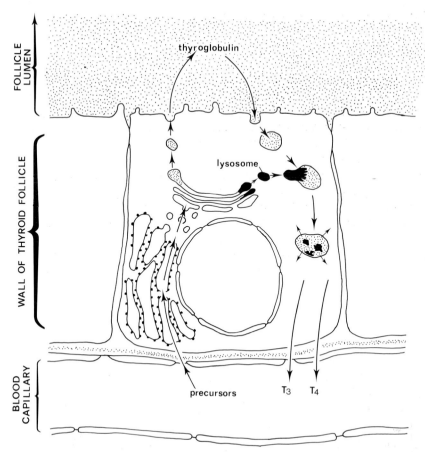

FOLLICLE LUMEN

WALL OF THYROID FOLLICLE

BLOOD CAPILLARY

thyroglobulin

lysosome

precursors   T₃   T₄

Figure 126. *Participation of lysosomes in the processing of thyroglobulin and the release of the thyroid hormones, T₃ and T₄.*

secretory granules that can go on to eventually release their content at the plasma membrane are once again produced. Crinophagy is probably most common in endocrine cells such as those of the anterior pituitary gland, where there is a constantly changing pattern in the stimulation and inhibition of secretion.

In thyroid follicular cells, lysosomal hydrolases may play a constructive role in processing secretory product. In the follicles of the thyroid gland (Figures 125 and 126), secretion is stored extracellularly as a large, iodine-containing glycoprotein, thyroglobulin; the active hormone principles secreted into the bloodstream are the iodinated amino acids tri-iodothyronine (T_3) and tetra-iodothyronine (thyroxine or T_4). Degradation of thyroglobulin to yield T_3 and T_4 occurs within the lysosomal system of the follicular cell; the thyroglobulin gains access to the lysosomes following its endocytotic uptake from the extracellular store. The iodinated amino acids produced then readily diffuse across both the lysosomal membrane and the follicular cell plasma membrane. Once degradation within the lysosome begins, it probably proceeds down to the level of the iodinated amino acids, and thus the primary control over the rate of thyroglobulin degradation must presumably be exercised over the rate at which it is taken into the cell from the extra-cellular store.

'Proteolysis' see
Proteases

Intracellular processing requiring 'proteolysis' is necessary for other secretory products (for example, in the conversion of proinsulin to in-sulin); however, since processing is usually concerned with producing a macromolecular end-product, it is unlikely to involve lysosomal elements.

The uptake and intracellular breakdown of low-density lipoprotein

Cholesterol. This is the most abundant steroid in animal tissues; it is a constituent of the plasma membrane and plasma lipoproteins, and is the precursor for steroid hormones and bile acids.

All cells require cholesterol, which they use primarily for the synthesis of their plasma membranes. However, although most cells are able to synthesize this molecule themselves, it seems that if an alternative, extracellular source is available, they will suppress endogenous produc-tion and draw in preference on the extracellular supply. In the blood, most cholesterol is carried within low-density lipoprotein (LDL) complexes, and cells are able to take up these complexes by absorptive endocytosis (see page 87). The uptake is specific and has been shown to depend upon the LDL complexes binding to specialized endocytosis-inducing receptors distributed in patches on the surface of the plasma membrane. Within the cytoplasm the pinched off, endocytotic vesicles enter the vacuolar system by fusing with one or more lysosomal elements, and within the secondary lysosomes produced the LDL complexes are degraded. The cholesterol released by the breakdown of these com-plexes is then transported to the surrounding cytoplasm, where it is either used directly for plasma membrane synthesis or modified (as a liquid crystal of cholesterol ester) for storage.

In the condition known as hypercholesterolaemia there is a marked and sometimes even complete loss of the LDL surface receptors. In severe conditions the blood level of LDL is elevated by ten-fold and, as a consequence, atherosclerosis and an early death usually occur. This and other evidence suggests that, while LDL uptake and degradation provides cells with an important source of cholesterol, it is probably of much greater significance in the overall control of the extracellular chole-sterol level.

The lysosomal membrane

Although there are probably proteolytic inhibitors in the cytoplasmic matrix of most cells, the integrity of the lysosomal membrane must be maintained if the cell is to be protected from the degradative activity of its own lysosomal enzymes. The lysosomal membrane is indeed less liable to disrupt than other cell membranes, and the slowness with which acid hydrolases leak from lysosomes in tissue homogenates is so well defined that 'latency' (i.e. a delay in the appearance of enzyme activity) is regarded as a diagnostic characteristic of a lysosomal enzyme. The reason the lysosomal membrane (and the lysosomal enzymes themselves, since most of them are glycoproteins) remains undegraded by the contained degradative enzymes has not been satisfactorily explained, although there is some evidence for a protective coating of highly charged glycoproteins on the inner surface of the membrane. Only under the unusual pathological circumstances when heterophagy leads to the incorporation of damaging agents, such as certain bacterial toxins, carcinogens or particles like asbestos and silica, are the limiting membranes of lysosomes known to become damaged and leaky. The uncontrolled release of acid hydrolases which then occurs can cause cell death. It is, nevertheless, worth emphasizing that this process is unusual and quite different from that occurring in the structural remodelling of a tissue; then, as described above, an orderly process of programmed cell death, relying upon autophagy, occurs.

Fusion amongst the various kinds of heterophagocytic and autophagocytic secondary lysosomes is common and probably random, and thus the convenient separation of the routes shown in the scheme outlined in Figure 122 should not be taken too literally. Lysosomes must also commonly fuse with other membrane-bound derivatives of the plasma membrane (e.g. heterophagosomes) and the Golgi complex (e.g. secretory granules), and clearly, because of the consequences of their contents mixing, this process must be selective. Identification of the membranes participating in these fusion processes presumably relies upon an ability to discriminate between their respective surface components. As yet, however, there is little direct information on the nature of these surface components, although it is known that some intracellular parasites, such as *Mycobacterium tuberculosis*, are able to modify the membranes of the phagosomes in which they are contained and prevent them from fusing with lysosomes.

In this context, the recent isolation and identification of the protein 'clathrin' should also be mentioned. This protein (mol. wt. about 180 000) is the major and perhaps only constituent of the 'fuzzy coat' which is commonly present on the cytoplasmic surface of small vesicles in the central nervous system. Clathrin is very likely to be a major component of similar coats on the cytoplasmic surface of both endocytotic invaginations (see page 85) and free cytoplasmic vesicles in many other cell types. It is believed that, in many cells, coated membranes tend not to fuse with lysosomal elements, so these 'protected' vesicles may provide a non-lysosomal intracellular pathway whereby extracellular proteins can be transported through the cell. The transport of maternal immunoglobulins across the placenta and across the intestinal mucosa in the newborn probably occurs via this kind of protected vesicle.

Specialized lysosomal elements

Even though the lysosomal population of any given cell type is typically heterogeneous in form, the enzyme complement of the lysosomes in most cells seems to vary only within narrow limits, and most appear to contain the same battery of approximately 60 different enzymes. There are, however, some lysosomal and lysosomal-like elements with a very distinctive enzyme content. For example, in the spermatozoon the head-piece contains a modified lysosomal structure called the 'acrosome', and it is the function of the enzymes within this membrane-bound vacuole to aid penetration into the oocyte at the time of fertilization. In the acrosome there is a predominant and characteristic acid protease known as 'acrosin'. Similar distinctive proteases occur in the lysosomes of blood platelets and in the azurophilic granules of granular leucocytes. The so-called 'specific' granules of the neutrophil granular leucocyte (i.e. those granules that are in the majority and give this kind of granulocyte its characteristic staining colour) do not contain acid hydrolase enzymes, but in their mode of formation from the Golgi complex and fusion with heterophagosomes they may be considered as a specialized component of the vacuolar system. These granules contain a disparate collection of compounds which include an alkaline phosphatase and bactericidal agents.

Lysosomes and disease

Although proteolytic enzymes have been implicated as causal factors in inflammatory conditions like arthritis and in malignancy, it is difficult to define precisely the role of lysosomes in these conditions. One of the most direct links between lysosomes and disease is demonstrated in the several genetic diseases which are related to single enzyme deficiencies. In these so-called 'storage diseases' the lack of a particular lysosomal hydrolase leads to the accumulation of substrate within the organelle. For example, in glycogen storage diseases, hypertrophied lysosomes distended with glycogen particles are a feature of liver parenchyma cells. Conditions resulting from deficiencies in sulphatases, phosphatases and lipases have all been described.

Of special interest for the treatment of these storage diseases is the finding that some cell types (e.g. fibroblasts and fixed macrophages in the liver) bear special receptors on their external surfaces that are capable of binding lysosomal enzymes and inducing their uptake by absorptive endocytosis. The physiological significance of this mechanism is unclear but in the treatment of patients with a storage disease due to a deficiency of a particular lysosomal enzyme it has provided a useful means of introducing the required hydrolases into enzyme deficient cells.

PEROXISOMES

Within the cells of the liver parenchyma and the proximal convoluted tubule of the nephron, small (300 to 1500 nm diameter), membrane-bound organelles called 'peroxisomes' are found (see Figure 103). They are fairly numerous (about one to every four mitochondria) and probably arise directly from the cisternae of the rough endoplasmic reticulum. They have no obvious association with any other organelle.

All peroxisomes contain several oxidative enzymes which yield hydrogen peroxide as a reaction product. The haemoprotein 'catalase' is also a characteristic constituent and may account for up to 40 per cent of their total protein. This enzyme is able to reduce hydrogen peroxide to water using either hydrogen peroxide itself or an alternative electron donor such as an aliphatic alcohol. As shown in Figure 127, the peroxidatic activity of the catalase allows the DAB reaction (page 40) to be used to identify peroxisomes.

Haemoprotein. Iron-containing protein.

Reduction. Any reaction in which an atom gains electrons.

The function of peroxisomes is at the present time obscure, since it is difficult to show that they have an essential role in the metabolism of either the liver or the kidney cell. In liver parenchymal cells they do, however, increase in number following the administration of certain chemical agents, such as salicylates. It may, perhaps, also prove significant that catalase levels are reported to be depressed in the livers of tumour-bearing animals.

SECRETORY GRANULES

Cells that synthesize proteins for export seem either to secrete them continuously at a slow rate in an unconcentrated form, or else they secrete them intermittently, storing them in the intervening, non-secretory periods as a concentrate within secretory granules. Fibroblasts, which synthesize and secrete collagen (the major protein of the intercellular matrix), are an example of cells that do not normally store their secretory product in secretory granules. Plasma cells (see Figure 97), although able to release their secretion at a rapid rate, are another. In contrast, secretory granules are a major feature in the cells of most glandular epithelia.

The number of secretory granules within a cell varies with the nature of its secretion and its secretory activity. In the pituitary, for example, a secretory cell may be quiescent for long periods of time and, in this storage condition, its cytoplasm is packed with secretory granules (Figure 128). In these cells, stimulation in the normal, physiological range is probably accompanied by the release of only a small proportion

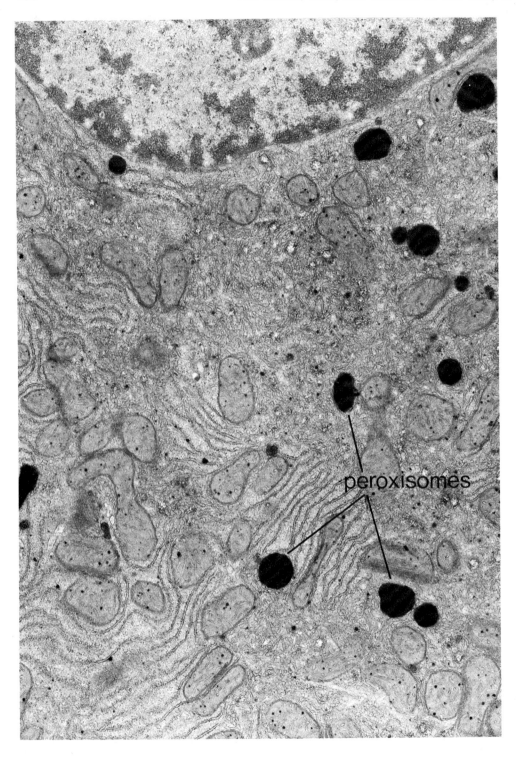

peroxisomes

Figure 127

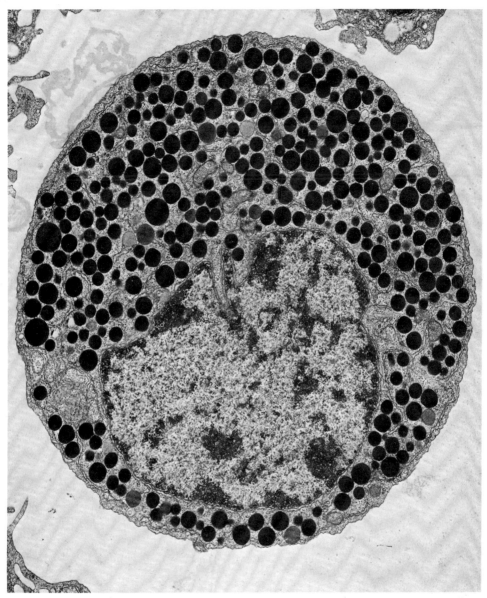

Figure 128. *Secretory granules distributed throughout the cytoplasm of a cell isolated from the pituitary gland. The dense population of granules and the few, sparse cisternal elements of rough endoplasmic reticulum indicate that this cell is in a quiescent, storage phase. The granules are known to contain growth hormone. They are all of the same size, and appear to vary in diameter simply because the section plane transects them at different levels. Magnification ×8000.*

Figure 127. *(Opposite.) Peroxisomes distributed throughout the cytoplasm of a liver parenchyma cell. Because of its peroxidatic activity, the catalase enzyme contained within the peroxisome yields an electron-opaque reaction product when treated with 3',3'-diaminobenzidine and osmium tetroxide. Under the appropriate conditions, this histochemical procedure can thus be used to stain peroxisomes while the form and distribution of the other organelles remain well preserved. Magnification ×12 500.*

Courtesy of H. D. Fahimi, Department of Anatomy, University of Heidelberg.

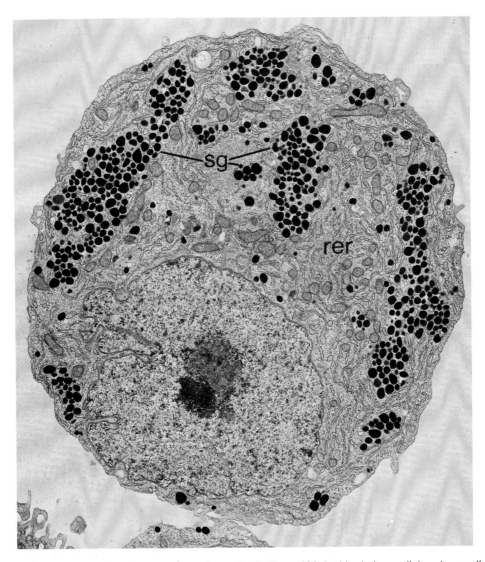

Figure 129. *A cell from the same preparation as that in Figure 128. In this pituitary cell there is a smaller population of secretory granules (sg), and the cisternal elements of the rough endoplasmic reticulum (rer) are more elaborately developed. Again, the sectioned profiles of the granules show a wide size range, although the population is homogeneous. The apparent discrepancy is emphasized here because the granules have an ellipsoid shape. Magnification ×8500.*

Exocrine

'Acinar cell' see *Acinus*

of the total secretory granule population, and it is thus only in cells where the store of secretory product has been depleted and there is a continuing demand for release that the organelles required for synthesis (the rough endoplasmic reticulum and Golgi complex) become a prominent feature (Figure 129). In exocrine cells, on the other hand, the demand for secretory product release is usually more frequent and makes heavier demands upon the cell content. For example, in the parotid gland, the acinar cells responsible for secreting the major enzymic constituents of saliva can be induced to release up to 80 per cent of their secretory product and are still able to replenish it within hours. In these cells, although there is normally a modest population of secretory granules, there is always an elaborate organization of cisternal elements concerned with further synthesis and packaging.

Secretory granules are always enveloped by a membrane, and their size and shape seem to be determined by the nature of their content. Consequently, the form of the secretory granules that contain specialized products, such as hormones, are very characteristic and they allow specific cell types to be identified with confidence. As a rule, although a secretory cell may synthesize and store many different secretory products (the acinar cells of the exocrine pancreas secrete at least a dozen different 'zymogens' – Figure 130), the secretory granule population is

Zymogen

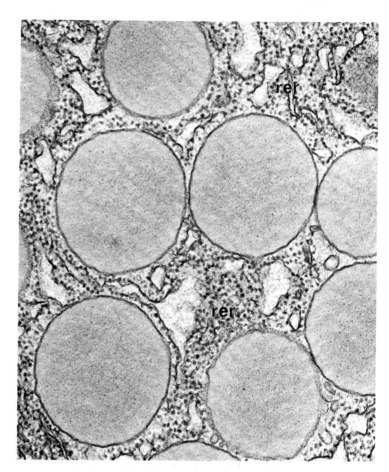

Figure 130. *A group of zymogen granules in an acinar cell of the exocrine pancreas. Although the content of these granules appears homogeneous, they each contain twelve or more different zymogens (the precursors of the enzyme constituents of pancreatic juice). Each granule is limited by a continuous membrane which is closely applied to the content. rer – rough endoplasmic reticulum. Magnification ×55 000.*
Courtesy of R. P. Bolender, Department of Biological Structure, University of Washington, Seattle.

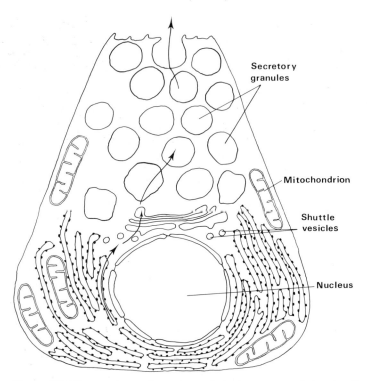

Secretory granules

Mitochondrion

Shuttle vesicles

Nucleus

Figure 131. *The secretory pathway in a polarized secretory cell showing the characteristic arrangement of the cisternal organelles and the secretory granules. From the time it is synthesized on the attached polyribosomes of the rough endoplasmic reticulum, secretory product is restricted entirely within the cisternal system. It is probably transported from the rough endoplasmic reticulum to the Golgi complex via small, 40 nm shuttle vesicles, and there it is concentrated and packaged into the secretory granules. Stimulation at the basal and lateral cell surfaces induces the release, by exocytosis, of secretory granule content across the apical boundary and into the duct. Most kinds of exocrine cell display this type of intracellular organization.*

homogeneous (i.e. all the types of zymogen are contained in each granule) and it is uncommon for there to be more than one kind of secretory granule in a cell. This presumably reflects the limited capacity of the cisternal systems to handle and keep separated large amounts of secretory product (see page 149). It also means that, regardless of the kind of stimulation that an individual cell receives, its response, in terms of the composition of its secretion, will always be the same.

In most cells, secretory granules arise directly from the releasing face of the Golgi complex (Figures 131 and 132). As described on page 166, this organelle, having processed and partially concentrated the secretory product, provides the enveloping membrane.

Although secretory granules provide the major storage compartment for protein secretion within cells, they should not be looked upon as simply inert containers. Within them a significant amount of secretory-product processing may occur. Thus, in the secretory cells in the adrenal medulla (which provide a rather extreme example), the granules contain

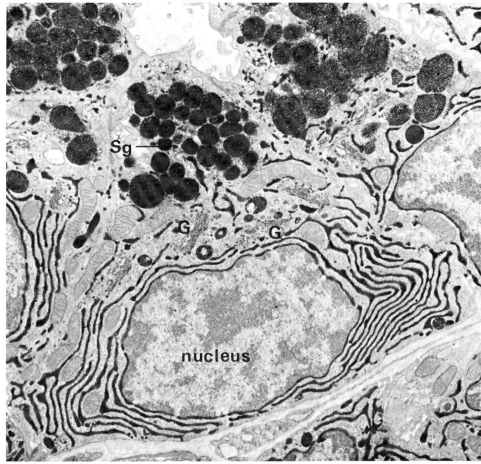

Figure 132. *The secretory pathway in an exocrine cell, outlined by specific staining of a secretory product. Amongst the secretory products of the lachrymal gland is a component (probably a bactericidal agent) with peroxidase activity, and by staining for this activity histochemically (see page 40) its distribution within the cell can be demonstrated. As shown in this micrograph, peroxidase secretory product is distributed throughout the rough endoplasmic reticulum (and the nuclear envelope), within some components in the Golgi complex (G), and within the mature secretory granules (sg) in the apical cytoplasm. The fine structural organization of these cells can be compared directly with that outlined in Figure 125. Magnification ×10 800.*
   Courtesy of V. Herzog and F. Miller, Department of Cell Biology, University of Munich.

the primary secretory product, catecholamine (21 per cent), together with the proteins and enzymes required for the final stages of catecholamine synthesis (35 per cent). They also contain significant amounts of adenosine triphosphate (15 per cent), and, in their enveloping membrane, there are the components of a complete electron transport chain (see page 200).

A major functional attribute of secretory granules is their ability to store, in high concentrations, biologically active substances of high potency. The beta cells of the islets of Langerhans, for example, store insulin within their secretory granules at a concentration equal to 15 per cent of the total dry weight of the cell.

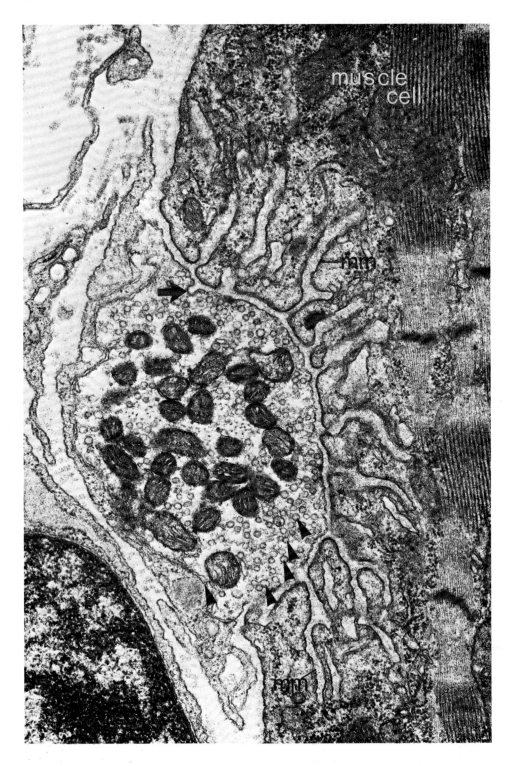

Figure 133

When the release of secretion is stimulated, secretory granules move to the cell surface and release their content by a process of exocytosis (Figure 133). In fusing with the plasma membrane, the granule membrane displays a high degree of selectivity. In polarized cells, such as those of the exocrine pancreas, the granules fuse only with the apical plasma membrane; presumably their site of fusion is determined by the molecular architecture of the inner surface of this membrane. The molecular basis of this recognition process and the agency responsible for inducing the exocytotic event remain to be identified. In most cells the entire process is complete within a fraction of a second.

Apical membrane

THE MITOCHONDRION

The mitochondrion is a membrane-bound organelle that lies free within the cytoplasmic matrix (Figure 134). Its most important functions are providing the cell's most widely used high-energy compound, adenosine triphosphate (ATP), and controlling the level of calcium ions in the cytoplasmic matrix. In cultured cells observed with phase contrast optics, mitochondria are seen to be actively mobile and display a perpetual wriggling and twisting motion. However, in cells with a high degree of internal organization (such as those of striated muscle and the distal convoluted tubule of the nephron, and in spermatozoa), they tend to be restricted within cytoplasmic compartments immediately adjacent to the sites of highest energy demand (see Figure 53).

Structure

The shape of mitochondria varies from spherical to thread-like, although in any given cell type they usually have a consistent form. They are always bounded by a smooth limiting membrane, and their internal organization, which is very characteristic, consists of two inner compartments separated by an inflected inner membrane (Figure 135). The inflections of the inner membrane are called 'cristae', and most typically (as, for example, in the acinar cells of the exocrine pancreas) they are

---

Figure 133. (*Opposite.*) *The release of the neurotransmitter, acetylcholine, from motor nerve endings may be regarded as a form of secretion. In this kind of nerve ending, which, as shown here, may innervate striated muscle fibres, the acetylcholine is stored in membrane-bound vesicles (small arrows). In response to an action potential propagated along the nerve axon, these vesicles fuse with the plasma membrane and release their content by exocytosis. The released neurotransmitter acts within the immediate vicinity of its site of release by binding to specific receptors on the external surface of the highly inflected muscle fibre membrane (mm).*

It is believed that the vesicular membrane contributed to the plasma membrane of the nerve ending by exocytosis is withdrawn from the surface for re-use by endocytosis. The endocytotic vesicles are thought to be of the coated variety; a typical example is indicated by the large arrow. Magnification ×32 000.

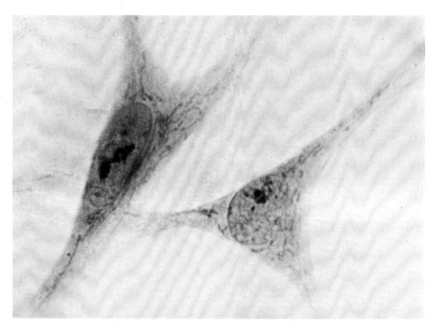

Figure 134. *A light micrograph of fibroblasts in tissue culture. Note the thread-like mitochondria distributed throughout the cytoplasm.*

shelf-like in form. In general, the total surface area of the cristae within a cell's mitochondria reflects the overall energy demand of the cell. Thus, for example, in cardiac muscle, where the ATP requirements for myofibril contraction may be especially high, the cristae are large and closely packed (Figure 136).

The membranes of the mitochondrion are about 6.5 nm thick, so they are slightly thinner than, for example, those of secretory granules or the plasma membrane. Nevertheless, in thin sections they show a typical three-leaflet structure. On the innermost surface of the inner cristal membrane there is an even distribution of closely packed 8.5 nm particles. *Negative staining* In negatively stained preparations of mitochondria, where these particles can be seen to good advantage, they appear to have a discrete, spherical head which is attached to the membrane by a short, narrow neck. The outer mitochondrial membrane is freely permeable to water and ions. The inner membrane, which forms the boundary to the intramitochondrial matrix, is, on the other hand, particularly impermeable. The transport of ions and metabolites across this inner barrier thus requires specific, energy-dependent carrier mechanisms.

Figure 135. *(Opposite.) A mitochondrion and rough endoplasmic reticulum in the cytoplasm of a secretory cell in the exocrine pancreas. Within the mitochondrion, the inner membrane forming the shelf-like cristae clearly divides the organelle into two compartments. The innermost compartment (the matrix) contains several prominent 'intramitochondrial granules'. Magnification ×80 000.*
Courtesy of K. R. Porter, Department of Molecular, Cellular and Developmental Biology, University of Colorado.

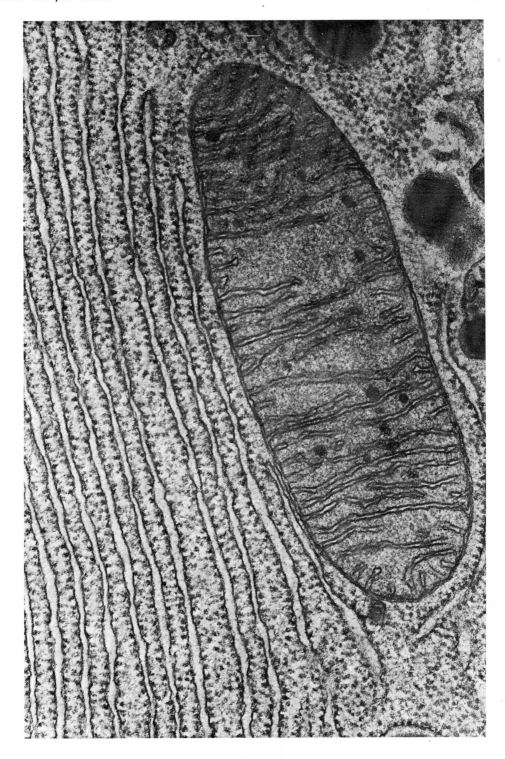

Figure 135

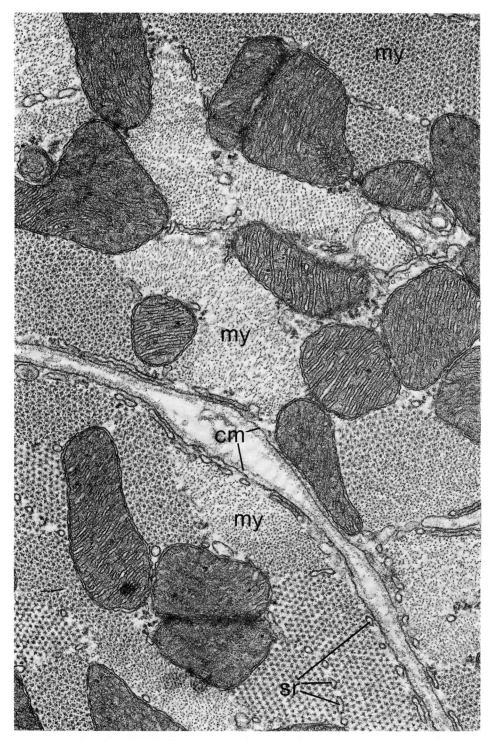

Figure 136. *A cross section of two adjacent cardiac muscle fibres to show the form of the mitochondria. These organelles lie within the cytoplasm between the bundles of myofilaments (my). Characteristically, they contain closely packed, shelf-like cristae. The plasma membrane is labelled 'cm'; the sarcoplasmic reticulum, 'sr'. Magnification ×40 000.*

Courtesy of C. Peracchia, Department of Physiology, University of Rochester.

The intramitochondrial matrix, which lies within the inner mitochondrial membrane, usually has a fine, fibrillar structure, and as a rule contains a variable number of dense, so-called 'intramitochondrial particles' (30 to 40 nm diameter). In the matrix, ribosomes (25 nm diameter) and fine strands of DNA have also been identified.

Functions

Oxidative phosphorylation

Mitochondria are the exclusive cellular location of the enzymes and cofactors required for the tricarboxylic acid (TCA) cycle (Formula 9).

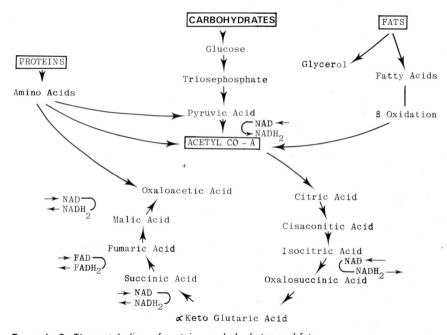

Formula 9. *The metabolism of proteins, carbohydrates and fats.*

This cycle oxidizes acetyl-coenzyme A (acetyl-CoA), a two-carbon unit which is produced mainly from pyruvate (the pyruvate being derived from the surrounding cytoplasmic matrix, where it is produced as a major end-product of glycolysis). The oxidation of coenzyme A in the TCA cycle is accompanied by the release of carbon dioxide and results in the transfer of hydrogen atoms to the coenzymes nicotinamide adenine dinucleotide (NAD^+) and flavin adenine dinucleotide (FAD) (Formula 9). When the reduced coenzymes are oxidized, their acquired hydrogen is split into protons (H^+) and electrons. The electrons are then transferred to an electron transport chain, which includes a sequence of compounds (NADH, NADH dehydrogenase, iron–sulphur protein, coenzyme Q

Oxidation. Includes any reaction in which an atom loses electrons, as, for example, when a ferrous ion (Fe^{++}) changes to a ferric ion (Fe^{+++}).

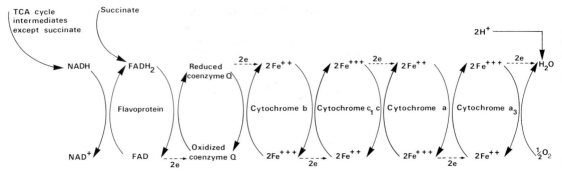

Formula 10. *The electron transport chain.*

[ubiquinone], and the cytochromes b, c_1, c, a and a_3) (see Formula 10). As the electrons are transferred along the chain, free energy is liberated, and each of these compounds is reduced in turn until, with the final reduction, oxygen is united with the H^+ ions to form water. In the transfer of electrons along the chain, three sites (I, II and III) have been identified at which the release of free energy is sufficient to generate a molecule of ATP from ADP and phosphate (Figure 137).

The process of oxidative phosphorylation, which includes the oxidation of pyruvate by the tricarboxylic acid cycle and then the phosphorylation of ADP by the electron transport chain, is carried out entirely within the mitochondrion. Most of the TCA cycle enzymes lie within the inner matrix, although at least one of them (succinic dehydrogenase) seems to be an integral component of the inner membrane. All of the components of the electron transport chain, on the other hand, are constituents of the inner membrane. The increased surface area provided by the inflections of the cristal membrane are important for the accommodation

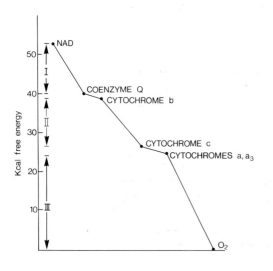

Figure 137. *The loss of free energy as electron pairs are transferred along the electron transport chain. The energy yielded at sites I, II and III is sufficient to generate a molecule of ATP from ADP and phosphate. Modified from A. L. Lehninger.*

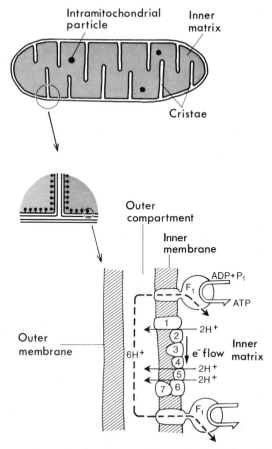

Figure 138. *The internal organization of the mitochondrion and its probable relationship to oxidative phosphorylation. H^+ yielded by oxidation (the TCA cycle, etc.) in the inner matrix are transported outwards, across the inner membrane, by the components of the electron transport chain (1, NADH dehydrogenase; 2, non-haem protein; 3 to 7, cytochromes b, c_1, c, a and a_3). Because of the impermeability of the inner membrane to passive transport by H^+, an ion gradient develops across the membrane, between the outer compartment and the inner matrix. The coupling mechanism of this membrane (which includes the so-called 'F$_1$ particles' on the inner surface) is then able to employ the potential energy contained within this gradient to phosphorylate ADP.*

of these components, for, although its protein content is high (80 per cent), only about 17 per cent of them are concerned with electron transport; the others are presumably related to the many highly specific active transport systems (e.g. for calcium – see below) that are also present in this membrane.

The components of the electron transport chain are distributed along the inner membrane of the mitochondrion in a regularly spaced array. The intermolecular relationships within these 'respiratory assemblies' are not entirely clear but some idea of their probable arrangement is given in Figure 138.

The significance of this arrangement in the capture of the free energy liberated by the transfer of electrons is most satisfactorily explained by the chemiosmotic hypothesis of Mitchell. According to this hypothesis, the arrangement of the components of the electron transport chain, together with the impermeability of the membrane in which they are located, is able to use the energy released by electron transfer to transport the H^+ ions emerging from the electron transport chain to the outer surface of the membrane. As a result an electrochemical gradient of H^+ ions (and a transmembrane potential) is established across the membrane between the outer surface and the matrix. It is by transducing the potential energy 'stored' in this gradient that the inner mitochondrial membrane is able to synthesize ATP.

This is essentially the reverse of the situation found in the erythrocyte membrane (see page 75), where the Na^+/K^+ ATPase maintains an ion gradient using energy derived from ATP breakdown. Supporting the idea that the potential energy of an ion gradient can be used to phosphorylate ADP is the finding that when, under the appropriate experimental conditions,* the Na^+/K^+ pump of the erythrocyte is made to reverse, and pump Na^+ inwards, it synthesizes rather than hydrolyses ATP.

The location and nature of the factors coupling the ion gradient to phosphorylation have not yet been clearly established, although it is probable that, like the components of the electron transport chain, they also will be found to be appropriately orientated within the inner membrane. Thus far, the only component identified with certainty is the so-called 'F_1 factor', and this is located within the 8.5 nm particles distributed along the inner surface of the membrane (see Figure 138).

Uptake of divalent cations

Mitochondria are able to concentrate calcium very efficiently; under appropriate conditions they can concentrate up to 25 per cent of their weight as calcium phosphate. The main sites of calcium storage are the 30 to 40 nm particles that lie within the intramitochondrial matrix, and *Osteoclast* in cells such as the bone-resorbing osteoclast, which are continuously exposed to an environment rich in calcium, these particles are particularly large and prominent. The mechanism of calcium concentration relies upon transport sites located within the inner mitochondrial membrane. It requires the expenditure of respiratory energy, and some indication of its importance to the cell (in regulating the level of intracellular calcium) is given by the fact that, in limiting conditions, the energy requirement for calcium transport is given priority over that required for ATP synthesis.

Control over the transient, local changes in the concentration of calcium ions that initiate and accompany a variety of cellular processes, such as muscle contraction, is probably exercised by specific pumping

---

* If K^+ ions are removed and Na^+ ions increased in the external medium surrounding the erythrocyte.

mechanisms such as those of the sarcoplasmic reticulum and the plasma membrane. This is because the level of calcium required to activate these pumping mechanisms is very low (it is in the micromolar range), while that required to stimulate Ca^{++} transport into the mitochondrion is some ten-fold higher. The role of the mitochondrion is therefore more likely to be in maintaining the general rather than the local calcium environment in the cytoplasmic matrix.

Mitochondria and lipid metabolism

Fatty acids derived from lipid storage deposits provide an important energy source, which can comprise up to 40 per cent of the total fuel requirement of the body. Their breakdown ('β-oxidation') to acetyl-CoA (which can then enter the TCA cycle) occurs in the matrix and on the inner membrane of the mitochondrion.   *β-oxidation*

    Mitochondria are also concerned with lipid synthesis, and, in particular, with the later stages of steroid hormone formation. These hormones are formed from cholesterol through a common intermediary, pregnenolone. This intermediary is transformed into progesterone (which is the progestational hormone synthesized and secreted by the placenta and corpus luteum), and this, in the appropriate tissue, is in turn a precursor for both the sex steroids (e.g. testosterone and oestrogen) and the adrenal corticosteroids (e.g. corticosterone). Some of the steps involved in these synthetic pathways (such as the conversion of pregnenolone to deoxycortisone) occur in the cisternae of the smooth endoplasmic reticulum, while others (such as the hydroxylation of deoxycortisone to corticosterone) occur in the mitochondria. In cells in which steroid synthesis is a major activity, the mitochondria and smooth endoplasmic reticulum are thus prominent and closely related features (Figure 139). Characteristically, although the reason is unclear, the mitochondria of these tissues contain large numbers of tubule-like cristae (Figure 140).

Heat production – non-shivering thermogenesis

If the newborn infant is exposed to cold, heat is generated in the brown fat-tissue located in the axilla and nape of the neck. The cells of this tissue are densely populated with mitochondria (it is the abundance of the chromoproteins in these mitochondria that gives the tissue its brown colour), and in this location their primary role appears to be thermogenesis. Whether or not the heat produced within the organelle arises from the free energy of the respiratory chain directly, or relies upon the use of ATP produced by a coupled oxidative phosphorylation, is not known.

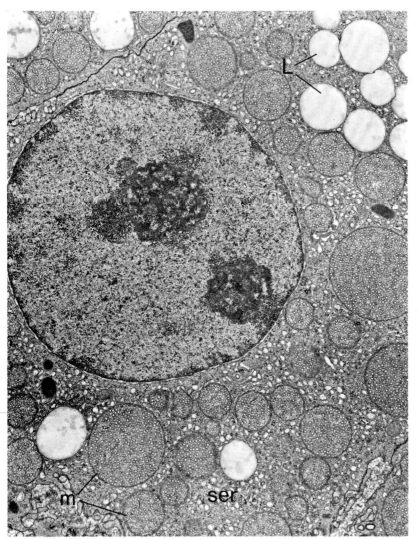

Figure 139. *A low-magnification electron micrograph of a cell from the adrenal cortex. This cell shows all the typical features of a steroid-secreting cell. The cytoplasm is packed with smooth endoplasmic reticulum (ser) and contains large, spherical mitochondria (m) and lipid droplets (L). The details of these features are shown more clearly in Figure 140. Magnification ×8500.*
  Courtesy of D. S. Friend and G. Brassil, Department of Pathology, University of California.

The origin or biogenesis of mitochondria

When a cell divides, the mitochondria become distributed amongst the progeny, and it is only during the subsequent growth phase of the cell that they, too, divide and increase in number. This individuality is a feature of mitochondrial biogenesis, and is emphasized by the fact that at no time in their development do they have a direct structural relationship with any other cytoplasmic organelle. Mitochondria are unique

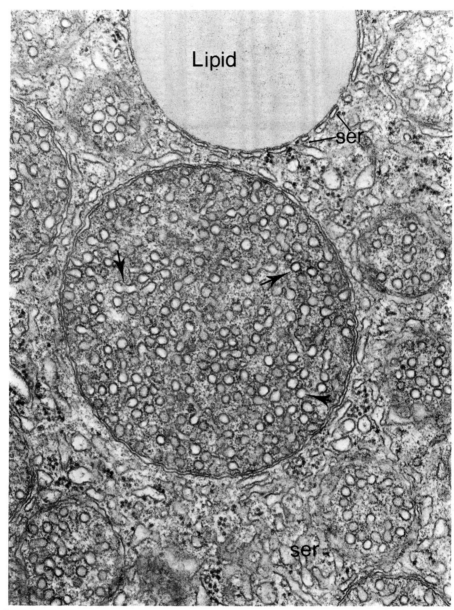

Figure 140. *A view of the cytoplasmic organization shown in Figure 139, at a higher magnification. Within the spherical mitochondrion there is an extensive development of the inner membrane to form tubular cristae (arrows). The smooth endoplasmic reticulum is also well developed, to form an extensive branching network (ser), which is often closely associated with large droplets of free lipid. Magnification × 23 400.*
  Courtesy of D. S. Friend and G. Brassil, Department of Pathology, University of California.

amongst the cytoplasmic organelles of mammalian cells in containing DNA, DNA-synthesizing enzymes (DNA polymerases), ribosomes, and all the factors required for ribosome-directed peptide synthesis. However, there are limits to their independence, since it is also clear that a proportion of mitochondrial proteins (including at least one of the cyto-chromes in the electron transport chain) are transcribed on nuclear DNA and translated on cytoplasmic ribosomes.

The evidence of semi-autonomy in mitochondria long ago gave rise to the suggestion that their presence in eukaryotic cells may have had its origin in a symbiotic relationship between an aerobic micro-organism and an anaerobic eukaryotic cell, the ancestor of the mitochondrion pro-viding ATP as its contribution to the relationship. Recent, more detailed examination lends additional support to this idea, since at the molecular level mitochondrial components have much in common with those of prokaryotes. Thus the DNA of mitochondria is circular and their ribo-somes are slightly smaller than their cytoplasmic counterparts. Further-more, protein synthesis on mitochondrial ribosomes is inhibited by the antibiotic, chloramphenicol, while it is insensitive to others, such as cycloheximide (a potent inhibitor of peptide-chain elongation on the cytoplasmic ribosome). All of these features are typical of prokaryotic systems. The dependence of the mitochondrion upon cytoplasmic trans-lation for some of its proteins need not be regarded as compelling evidence against its prokaryotic origin, since it may simply reflect a depen-dence (accompanied by a gradual erosion of the mitochondrial genome) that has grown since the original symbiotic relationship was established. The available evidence, although circumstantial, thus strongly favours a prokaryotic origin for the mitochondrion.

7

THE CYTOPLASMIC MATRIX

The 'ground cytoplasm' or 'cytoplasmic matrix', in which the membrane-bound organelles and cisternal elements lie, comprises both highly organized, structural components and a soluble phase. In actively mobile cells in culture, rapid, generalized transitions between a viscous phase and a labile, fluid phase are a major feature of the matrix. Within stationary cells, local variations in phase, giving rise to localized moving streams and avenues of protoplasm, are frequently observed. The molecular mechanisms underlying these changes in fluidity are unclear, although it is probable that changes in the degree of polymerization of the structural components of the matrix are an integral part of them.

In the centrifugal fractionation of cell homogenates (see Figure 30), the soluble phase of the cytoplasm is represented by the unsedimentable high-speed supernatant; within this fraction is found the major portion of the cell's metabolic machinery. Thus, in addition to the polymeric structural components of the matrix, it is clear from the analysis of these supernatants that the cytoplasmic matrix contains most of the cell's elemental components, such as water, ions and dissolved gases, as well as most of the metabolites of its major metabolic pathways. In the matrix, for example, are all of the enzymes and substrates required for glucose metabolism, together with their immediate energy requirements.

FORMED STORAGE PRODUCTS—GLYCOGEN AND LIPID

Macromolecular storage products, in the form of glycogen particles and lipid droplets, lie free within the cytoplasmic matrix.

Glycogen particles are prominent features of cells such as liver parenchyma (see Figure 88) and muscle cells (see Figure 105), which synthesize and store them. For light microscopy, glycogen aggregates are best demonstrated with the periodic acid—Schiff (PAS) method (see Figure 23), while, in the electron microscope, they can be readily identified as groups of small, discrete particles that stain strongly with lead salts. In the electron microscope, the smaller glycogen particles occur as fine, 3 to 20 nm rods, while the larger, so-called 'alpha' particles (the usual form of glycogen seen in liver parenchyma cells) have a coarse, irregular outline and may be up to 150 nm in diameter.

Cytoplasmic lipid, the alternative energy store to glycogen in cells, usually occurs as free, spherical droplets that may vary considerably in size. A large number of small lipid spheres, rather than a few large masses, usually indicates active use and turnover. Like glycogen, lipid droplets are a common feature of the cytoplasm of liver parenchyma cells, and in these cells the amount and distribution of droplets vary dramatically with changes in the physiological and pathological state of the organ.

In the cells of the adrenal cortex (Figure 139), corpus luteum, and other steroid-secreting tissues, lipid droplets represent a store of steroid precursor, rather than the hormone itself. Active secretion in these cells is typified by large numbers of small lipid droplets, and often the population of lipid droplets is so dense that it reduces the remainder of the cytoplasmic matrix to a network of fine, attenuated strands. In paraffin sections of this tissue prepared for light microscopy, the embedding procedure extracts the lipid, so the network of cytoplasm that remains has a characteristic 'foamy' appearance.

STRUCTURAL COMPONENTS OF THE CYTOPLASMIC MATRIX

The organized, structural components of the cytoplasmic matrix exist as variously developed networks and linear arrays of fine fibrils and tubules. At the present time our understanding of these structures derives primarily from a few specialized systems. However, since this is an area of especially active research interest, our knowledge will undoubtedly become much extended in the near future.

MICROFILAMENTS AND MUSCLE CONTRACTION

Striated muscle

The organization of the microfibrillar components concerned with force generation reaches its highest level in striated muscle cells (Figure 141). In these cells it is well established that in the appropriate ionic environment, and using energy derived from ATP, thick ('myosin') and thin ('actin') filaments can associate to generate contractile force. To date, this is the only interaction between microfilamentous structures that has been shown, for certain, to be capable of force generation.

Portion of a 'RED' muscle fibre

Portion of a 'WHITE' muscle fibre

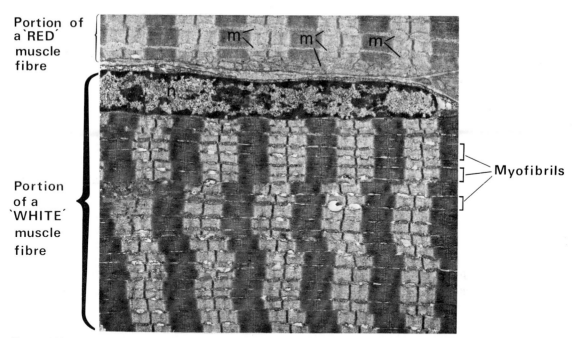

Myofibrils

Figure 141. *An electron micrograph of relaxed striated muscle fibres. Each fibre is an elongated cell (1 to 4 mm long, 10 to 40 µm diameter) containing many peripheral nuclei. In this micrograph, portions of two fibres are shown.*

In the lower fibre, a nucleus (n) and a number of myofibrils, with their characteristic striations, are seen in longitudinal section. This fibre is classified as a 'white' fibre because it has a large diameter and relatively few mitochondria.

In the upper fibre there are more mitochondria (m), and typically they lie in series, parallel to the myofibrils. This fibre is thus classified as a 'red' fibre. Magnification ×6000.

The molecular details of the interaction between the thick and thin filaments are embodied in the 'sliding filament' model of Hanson and Huxley. This model was initially put forward to account for the mechanism of contraction in skeletal muscle fibres, but it also accounts satisfactorily for the arrangement of thick and thin filaments in cardiac muscle cells. With modification it may also be applied to similar arrangements in smooth muscle and (almost certainly) to the contractile mechanisms of many non-muscle cell types. In outline, the arrangement and interaction of the thick and thin filaments in skeletal muscle are as follows.

The arrangement of the myofilaments

The contractile unit is the 'sarcomere'. Each sarcomere is delineated by 'Z-lines' and is repeated in series and parallel along the length of the muscle fibre (Figure 142). The thin filaments are anchored at one end in the Z-line material; at their free ends they interdigitate in a double hexagonal array with the thick filaments (Figures 143 and 144). In sections across the myofilament bundle, each end of a thick filament is thus seen to be associated with a group of six thin filaments (Figure 145).

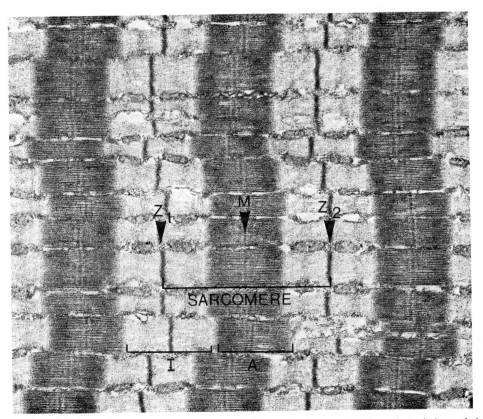

Figure 142. *A relaxed striated muscle fibre cut in longitudinal section to show the striations of the myofibrils. The myofibrils (1 to 3 μm diameter) are subdivided by their Z-lines into sarcomeres (Z_1 to Z_2), and they lie in register across the width of the fibre. Within each sarcomere, the striations indicate the distribution of the myofilaments. The I- (isotropic) bands contain the Z-lines with their attached thin (actin-containing) filaments, while the A- (anisotropic) bands contain not only the M-lines, with their attached thick (myosin-containing) filaments, but also the ends of the interdigitating thin filaments.*

On contraction, the thin filaments slide between the thick filaments, so that, although the width of the A-band remains unchanged, that of the I-band narrows. Magnification × 15 000.

'I-band' and 'A-band' see *Isotropic*

The thick filament (130 nm long, 12 to 15 nm diameter) consists mainly of a bundle of myosin molecules. Each individual molecule (mol. wt. 500 000) is shaped like a double-headed golf club (Figures 146 and 147) with a hinged shaft. Within each bundle, the myosin molecules are arranged in a bipolar fashion with their protruding club-like heads directed towards the ends and their shafts directed towards the middle. A myosin bundle, therefore, has a central, bare zone and an array of protruding heads of opposite polarity at each end (Figure 143).

Figure 144. *(Opposite.) A view at high magnification of contracted myofibrils from striated muscle. In this longitudinal section, the length of the sarcomeres is defined by the Z-lines. The thin filaments (in this section plane there are two between each thick filament — for clarification refer to Figure 145) reach between the thick filaments, almost to the M-line. For the area outlined in the box refer to Figure 146.*

Courtesy of H. E. Huxley, MRC Laboratory of Molecular Biology, Cambridge.

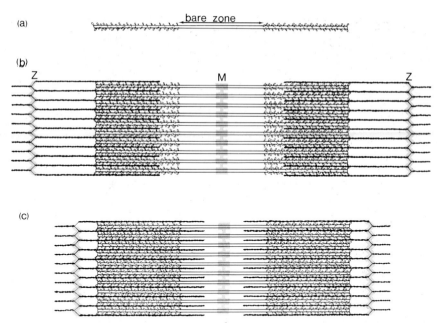

Figure 143. *The force-generating components of the contractile unit in striated muscle. Shown in (a) is a single thick filament to illustrate the bipolar distribution of the myosin heads and the central bare zone that contains the shafts (or tail portions) of these molecules.*

In (b), the thick and thin filaments partially overlap; this is the relaxed position. In (c), there has been an inward displacement of the thin filaments, which has resulted in a more extensive overlap and a corresponding reduction (contraction) in the length of the sarcomere.

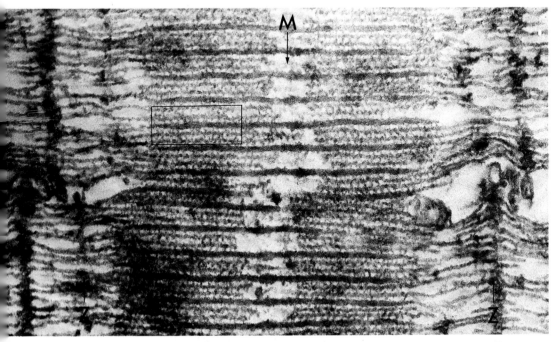

Figure 144

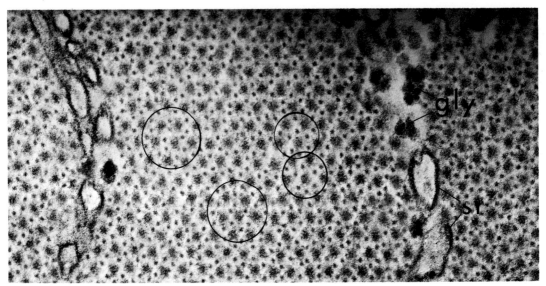

Figure 145. *A cross section of myofibrils in striated muscle to show the disposition of the thick and thin filaments. gly – glycogen; sr – sarcoplasmic reticulum.*
   Courtesy of H. E. Huxley, MRC Laboratory of Molecular Biology, Cambridge.

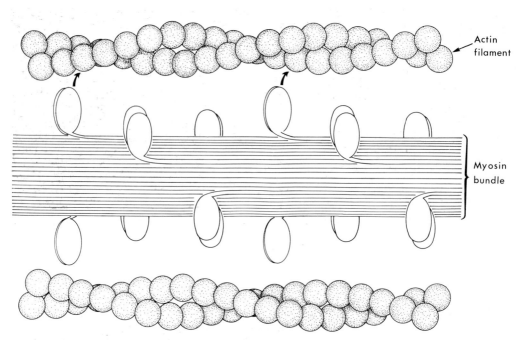

Figure 146. *A diagram showing the relative positions of the myosin and actin molecules as they might be arranged in the thick and thin filaments outlined in Figure 144. The double helix of the actin filament has 13 G-actin molecules (diameter about 5.6 nm) per turn. The heads of the myosin molecules are arrayed in six rows. There is a 120° displacement between each row of heads and a 43 nm periodicity between the heads in each row. It is believed that a hinge region on the shaft of the myosin molecule allows the myosin head to be displaced (about 15 nm) in order to bind to specific sites on the globular subunits of the actin filament (see arrows).*

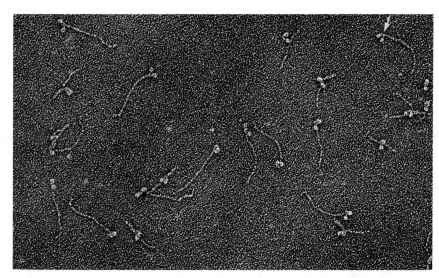

Figure 147. *A high-resolution electron micrograph showing individual, platinum-shadowed myosin molecules. In each of them the long shaft region bears a double-headed end. In some instances (arrow) the separation of the two heads and their angle of attachment to the shaft are shown very clearly. Magnification ×104 000. Courtesy of A. Elliot and G. Offer, Department of Biophysics, University of London, King's College.*

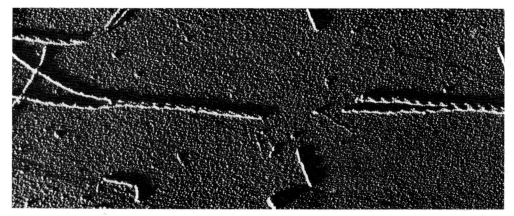

Figure 148. *A high-resolution micrograph showing isolated filaments of F-actin. There is a clear indication of their helical form. Magnification ×104 000. Courtesy of R. V. Rice, The Mellon Institute, Pittsburgh.*

The major component of the thin filaments is F-actin, a filamentous protein composed of helical polymeric chains of the globular protein, G-actin (5 nm diameter, mol. wt. 42 000) (Figures 146 and 148). F-actin chains also have a linear polarity, and although the molecular basis of the polarity is unknown, its direction can be demonstrated using 'heavy meromyosin'. This enzymic digest consists of only the head portions of the myosin molecule, and in the electron microscope it is seen

Filamentous protein

Globular protein

Polarity. With a preferred direction.

to coat the thin filaments with arrow-shaped fragments in a linearly polarized array. It is important to note that all of the actin filaments inserting into one side of a Z-line have the same polarity, and all of the filaments inserting into the other side (and those inserting into the Z-line at the other end of the same sarcomere) have the opposite polarity.

In addition to F-actin, thin filaments also contain the regulatory proteins, tropomyosin and troponin; these will be discussed below.

The myosin – actin interaction

By virtue of their hinged shafts, the heads of the myosin molecules can move towards and bind with the globular actin subunits of the surrounding thin filaments (Figure 146). Contractile force is generated by the ratchet-like attachment, detachment and reattachment of these myosin heads; this, because of the preferred polarity in the arrangement of the two groups of interacting molecules, draws the thin filaments inwards towards the middle of the thick filament. Because the thin filaments are anchored in the Z-lines, the length of each contractile unit (the sarcomere) is thus reduced.

On relaxation, all of the myosin heads become detached and the two bundles of filaments slide back to their original position of partial overlap.

Energy requirements

The energy required for the cyclic association of the myosin heads with the actin filaments is derived by splitting ATP to form ADP and phosphate. In each cycle, each myosin head acts as an ATPase and splits one molecule of ATP. In binding to the actin, the ATPase activity of the myosin is stimulated and the rate of ATP breakdown increases. However, unless it can bind another molecule of ATP, the myosin is then unable to dissociate from the actin in order to progress through another cycle of reattachment. It is for this reason that, if the ATP level becomes limiting, a muscle fibre remains permanently contracted and is said to be in 'rigor' (as in rigor mortis).

Because there are only modest stores of ATP in striated muscle fibres, continued contraction must rely upon its continued replenishment by mitochondria. In fibres able to sustain repeated contraction (the so-called 'red' fibres of skeletal muscles), these organelles are arranged in series alongside the myofibrils (Figure 141).

Regulation of the actin–myosin interaction by calcium

In addition to the long, helical filaments of F-actin, the thin filaments of striated muscle contain the regulatory proteins, tropomyosin and troponin. Tropomyosin molecules (mol. wt. 70 000) are long and thin (40×

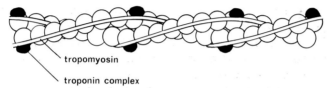

tropomyosin

troponin complex

Figure 149. *The probable arrangement of the tropomyosin and troponin complex components along the F-actin helix. After S. V. Perry.*

2 nm), and lie end-to-end alongside the groove of the actin helix. Each tropomyosin molecule is associated with seven actin monomers and, in a relaxed muscle, it probably covers (or otherwise blocks) the myosin-binding sites on these molecules (Figure 149).

Each troponin protein complex includes three globular subunits: troponin C (mol. wt. 18 000), troponin I (mol. wt. 22 000), and troponin T (mol. wt. 38 000). Troponin C is able to bind calcium ions, while troponin I can induce tropomyosin to move and reveal the myosin-binding sites on the actin filament. The function of troponin T is uncertain but it is probably responsible for the attachment of the troponin complex to tropomyosin.

Regulation of contraction depends upon the availability of calcium. On stimulation, when calcium is released from the sarcoplasmic reticulum, its concentration in the environment surrounding the myofilaments increases from less than 1.0 to about 10 micromolar (see page 157). Under these circumstances, calcium binds to troponin C and induces a change both in the conformation of troponin I and in the relationship of the troponin complex as a whole. This alteration causes the tropomyosin to move, and the ATP-activated myosin heads can now bind to the actin (Figure 150).

At post-stimulation, when the sarcoplasmic reticulum re-accumulates the calcium, the sequence is reversed: the troponin complex is deprived of its calcium, and the tropomyosin returns to its former position, preventing the myosin—actin interaction.

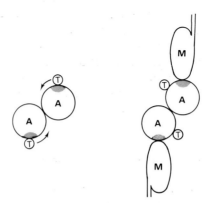

Figure 150. *A model indicating in 'end on' view the movement of the tropomyosin filament (T) which allows the myosin head (M) to bind to the actin subunit (A). After J. M. Squires.*

The dynamism of this calcium-regulated process is very impressive; in cardiac muscle, for example, the influx and withdrawal of calcium occurs between three and four times a second.

Although there is no doubt that calcium regulation of the troponin–tropomyosin complex represents the primary mechanism for the control of contraction in striated muscle, it is nevertheless very probable that additional regulatory systems are also involved. One of these systems may depend on the requirement of actin–bound myosin for ATP, which is related to the rigor condition mentioned above. Another may include a direct calcium regulation of the myosin molecule similar to that found in smooth muscle (see below). The nature and interrelationship of these systems, and their relative importance to the different kinds of stimulation (i.e. nervous and hormonal), remain to be clarified.

Additional components

Besides the primary force-generating components (actin and myosin) and the regulatory components of the troponin–tropomyosin complex, several other protein components have been identified in striated muscles. One of these is alpha actinin, a small, rod-like molecule (mol. wt. 200 000) that is a prominent component of the Z-line and has a regular, periodic distribution along the myofibril. Another is desmin, a protein (mol. wt. 50 000) that is believed to be a major constituent of the 10 nm fibrils present in smooth muscle cells (see below). Since both alpha actinin and desmin are found in the Z-line (and also in the intercalated discs of cardiac muscle), it is believed that they are primarily concerned with the anchorage of the thin, actin-containing myofilaments.

In the middle of each sarcomere the thick filaments of each myofibril are held together at the 'M-line'; this region contains a distinctive, but as yet uncharacterized protein, the 'M-protein'. A second, linearly arranged protein component, known as the 'C-protein', has also been identified. This component lies parallel to the M-line at intervals coincident with the rows of myosin heads on the thick myofilament. As yet, the functions of these thick filament components are unknown.

Smooth muscle

In smooth muscle fibres the contractile unit is the cell; a distinct, organized array of myofilaments forming a sarcomere-like arrangement does not exist. Nevertheless, thin and thick filamentous components can be identified (Figure 151), and force generation in these cells depends largely, if not entirely, upon the interaction between actin and myosin. The bundles of thin filaments attach to the cell surface via plaque-like condensations (dense bodies – see Figures 152 and 153) and extend into the cell interior. They also probably attach to similar densities lying in the cytoplasmic matrix within the body of the cell.

Plaque. A patch or spot.

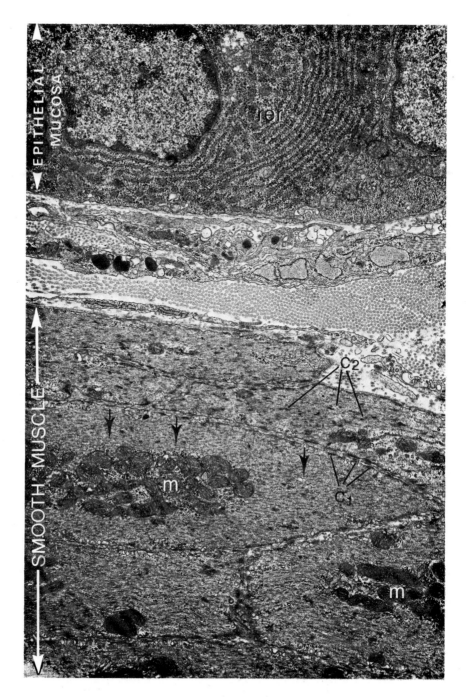

Figure 151. *A low-magnification view of the wall of the intestine, showing the base of the epithelial mucosa and the edge of the innermost smooth muscle layer. The ellipsoidal smooth muscle cells contain aggregations of mitochondria (m) and a dense mass of myofilaments. Amongst the myofilaments (most of which are below the limit of resolution in this micrograph) are a large number of short, dense filaments (arrows) orientated in the long axis of the cell. These 12 nm diameter filaments presumably represent the thick, myosin-containing components. Below the plasma membrane of the cells there are regularly arranged condensations (labelled 'C₁'). There are similar condensations in the body of the cell (labelled 'C₂').*

In the cell at the top of the micrograph, the parallel array of rough endoplasmic reticulum (rer) is typical of a cell synthesizing protein for secretion. Magnification ×11 700.

Courtesy of C. Peracchia, Department of Physiology, University of Rochester.

217

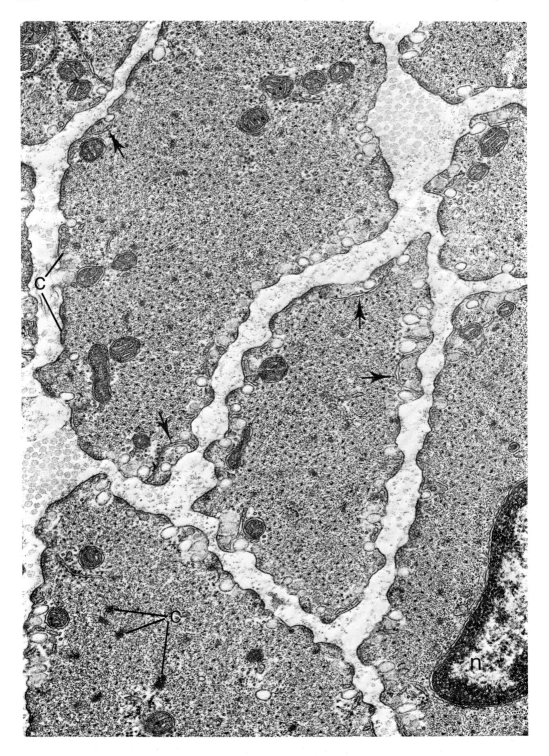

Figure 152

The thin filaments contain F-actin, tropomyosin, and at least some of the components of the troponin complex. Myosin is also present in smooth muscle, but relative to the amount of actin there is proportionally much less than in striated muscle cells. The myosin molecules are believed to be located within discrete thick filaments (about 2000 nm × 15 nm diameter), but with conventional fixation procedures these structures do not seem to be as stable as those of striated muscle.

In smooth muscle, calcium ions are again the primary intracellular signal for the initiation of contraction. At the present time, however, the best evidence suggests that they interact primarily with the myosin rather than the actin. It has been shown that calcium ions are able to induce a change (a phosphorylation) in the myosin molecule, and that *Phosphorylation* this alteration precedes the activation of the myosin as an ATPase. Thus, only when it is activated by calcium can the myosin hydrolyse ATP and bind to actin. This mechanism is independent of any regulation by the thin filaments. Although it is known that tropomyosin is associated with the actin of the thin filaments, it has not yet been shown that they include a responsive troponin complex. Whether or not calcium binds and regulates the activity of the actin in the thin filaments is thus unknown.

In addition to the thin (actin) and thick (myosin) filaments there is a third kind of filament in smooth muscle. These structures are known as 'intermediate filaments' and are typically 10 nm in diameter. They insert into the plaque-like condensations on the plasma membrane and, although it has not been shown directly, they probably also insert into the dense plaques that lie within the body of the cell (Figure 154). It seems likely that the role of these filaments in smooth muscle is primarily supportive and concerned with providing a framework for the attachment of the force-generating components. Desmin (mol. wt. 50 000) has been extracted from the intermediate filaments, and by immunocytochemical methods has been shown to be a major constituent of the plasma membrane condensations which serve as points of insertion for the thin (actin) filaments. As mentioned above, this protein has also been shown to be a component of the Z-lines, which provide anchorage for the thin filaments of striated muscle fibres.

Recently, another filamentous protein has been isolated from smooth muscle. Known as 'filamin', it is composed of two subunits (each of mol. wt. 250 000) and, like tropomyosin, it is not only associated with

---

Figure 152. (*Opposite.*) *A low-magnification view of smooth muscle fibres in cross section.*
The plasma membrane of these fibres is frequently invaginated to form 'surface vesicles'. However, although these are a prominent and characteristic feature of all smooth muscle fibres, their functional significance is unknown. Between the 'surface vesicles', on the cytoplasmic side of the plasma membrane are 'dense bodies' (C), plaque-like condensations of fibrous material. As seen in the cell at lower left, similar condensations also occur within the body of these cells. In both situations they are believed to provide attachment sites for the thin, actin-containing myofilaments.
The cytoplasm in each smooth muscle fibre also contains a dense population of thick and thin myofilaments, mitochondria and, at the periphery, occasional cisternae of smooth endoplasmic (or sarcoplasmic) reticulum (arrows); n — nucleus. Magnification × 34 000.
Courtesy of A. P. Somlyo, Pennsylvania Muscle Institute of the University of Pennsylvania.

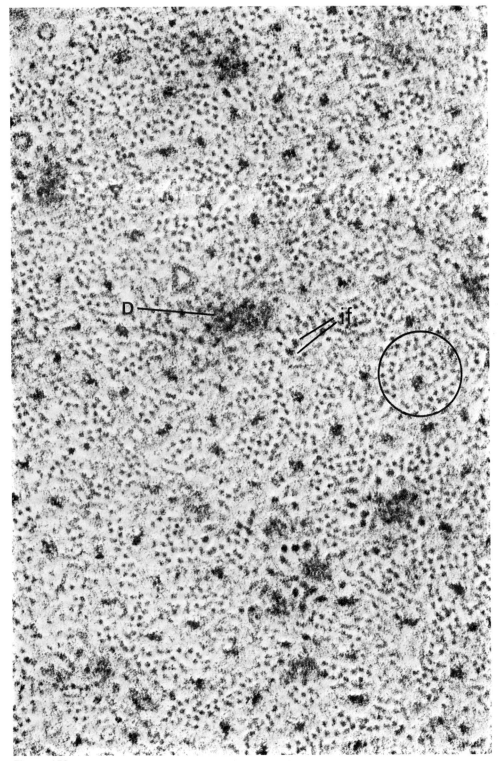

Figure 153

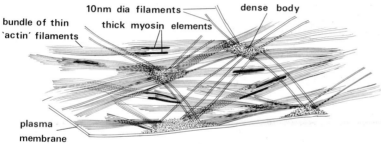

Figure 154. *A reconstruction of some of the interrelationships that may exist between the major filamentous components in the smooth muscle cell. The 10 nm diameter filaments are shown inserting into the dense bodies that lie in the body of the cell and at the plasma membrane. This arrangement provides a framework for the attachment of the bundles of 7 to 8 nm diameter actin filaments. At their free ends, the actin filaments are associated with thick, myosin-containing filaments. Contractile force generated by the interaction of the filaments containing actin and myosin, and transmitted via the inelastic 10 nm filaments, would cause the cell to contract along its long axis.*

actin but is present in consistent proportion with the actin content. Like myosin, and unlike the other filamentous proteins of muscle, there is some evidence that filamin can interact with actin to generate contractile force. Of particular interest is the observation that although this protein is also present in non-muscle cells (see below), it has not been found in striated muscle.

MICROFILAMENTS AND MUSCLE PROTEINS IN NON-MUSCLE CELLS

Contractile mechanisms are the basis of movement in many cellular processes. These processes may involve the whole cell, as in the locomotion of freely mobile cells, or they may involve local manipulations of the cell surface, as in cell division (at cytokinesis) or phagocytosis. Presumably, intracellular events such as the shuttling activity of the vesicular components of the vacuolar system and the movement of chromosomes prior to cell division also depend upon the participation of similar force-generating components. Primarily through studies that have exploited the specificity and exquisite sensitivity of the immunofluorescence technique, it has recently become clear that the filamentous components involved in many of these events are the same as those of muscle cells.

Figure 153. *(Opposite.) A high-resolution electron micrograph of a smooth muscle cut in transverse section to show the distribution of the myofilaments. As shown by the encircled region, the 'thick' (about 15 nm diameter) filaments are each surrounded by an inner halo of amorphous material and an outer array of 'thin' (5 to 8 nm) actin filaments. There is a thin to thick filament ratio of approximately 15 to 1. Dense bodies (D) are surrounded by intermediate (10 nm diameter) filaments (if). It is believed that both intermediate and thin filaments insert into these condensations. Magnification ×220 000.*
Courtesy of A. P. Somlyo, Pennsylvania Muscle Institute of the University of Pennsylvania.

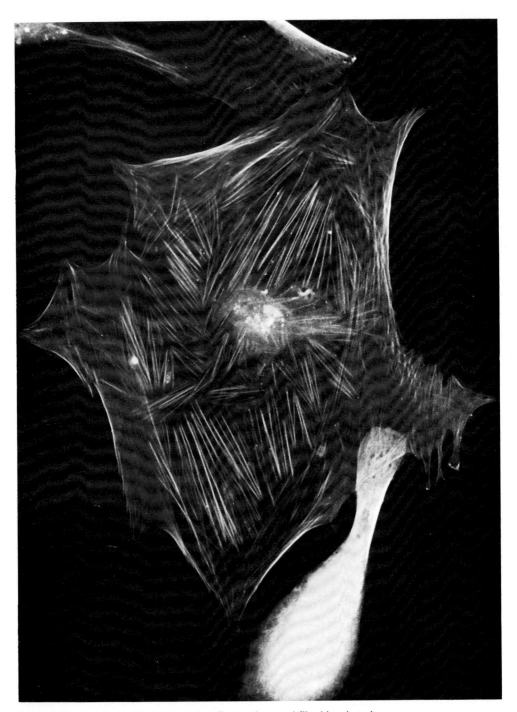

Figure 155. *The distribution of actin in a flattened, spread fibroblast in culture.*

The immunofluorescence, which depends upon an antibody directed against actin, shows groups of straight, so-called 'stress' fibres within the body of the flattened cell. At the cell periphery, the distribution of the actin components is more sheet-like. The brightly fluorescent structure to the lower right represents a portion of a thicker, rounded cell. Magnification ×320.

Courtesy of E. Lazarides, Department of Molecular, Cellular and Developmental Biology, University of Colorado.

At the present time a full account of their distribution and relative abundance cannot be given, but it is already apparent that in many respects the contractile mechanisms of non-muscle cells are best compared with those of smooth rather than striated muscle.

Actin in non-muscle cells

Actin is a major component of most non-muscle systems and in many cell types accounts for well over 10 per cent of the total cell protein. Immunocytochemical and electron microscope studies show that the distribution of filamentous actin in non-muscle cells is variable; sometimes it is organized into thick bundles, but more usually it forms a loose, finely fibrous network.

In flattened, immobile cells in culture, actin filaments are usually organized as straight coarse bundles that radiate throughout the cell. These filamentous bundles are called 'stress' or 'sheath' fibres and they are most clearly demonstrated by high-voltage electron microscopy (see Figure 97). Immunofluorescence (Figure 155) shows that stress fibres contain actin (and myosin — see below) but, interestingly, when their polarity is demonstrated by heavy meromyosin binding, their constituent filaments show no preferred direction of alignment.

In moving cells, stress fibres lie parallel to the direction of movement and as the cell changes direction they change their alignment accordingly (Figure 156). Since these filamentous bundles do not continue into the ruffled border area at the leading edge of the moving cell, it seems unlikely that they are concerned with the extended, forward phase of movement (see page 229). More probably, they are responsible for 'bringing up the rear', when the extended cell contracts (see Figure 162).

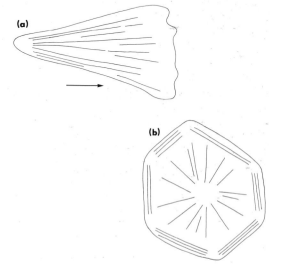

Figure 156. *The arrangement of stress fibres in cultured cells. In actively mobile cells (a) the filaments are arranged in the long axis of the cell parallel to the direction of movement. When the cell changes direction the filaments become realigned.*

When cells in suspension are added to a culture dish they usually attach and then spread on the substratum (b). At this time (see also Figure 155) peripheral stress fibres appear. While some of these fibres lie parallel to the cell perimeter, others radiate outwards from the centre. After N. K. Wessels.

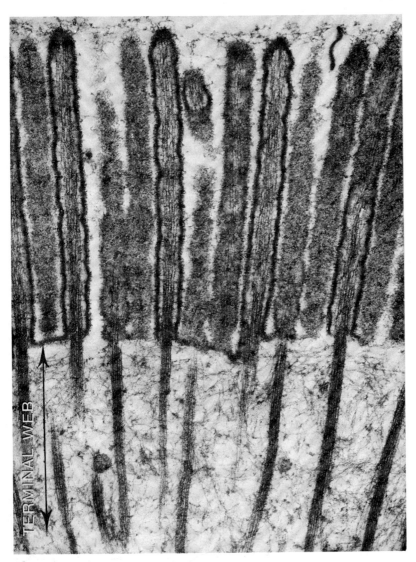

Figure 157. *A microvillous (brush) border isolated from a cell lining the small intestine, showing the form and distribution of the cytoplasmic microfilaments.*

The bundle of filaments within the core of each microvillus extends from the tip, where it is attached, down into the complex of filaments within the apical cytoplasm (the terminal web). The major component of these bundles of filaments is F-actin, and (using heavy meromyosin binding) it has been shown that its polarity is always directed towards the terminal web (that is, as in striated muscle, away from the site of attachment and towards the region of force generation).

Amongst the population of filaments within the terminal web there are other varieties of microfilament, which probably include, in addition to actin, both myosin and intermediate filaments. An arrangement in which these components may interact to generate contractile force, and which is consistent with the polarity of the actin filaments, is shown in Figure 159. Magnification ×60 800.

Courtesy of M. S. Mooseker and L. G. Tilney, Department of Anatomy, Harvard Medical School, Boston, and Department of Biology, University of Pennsylvania.

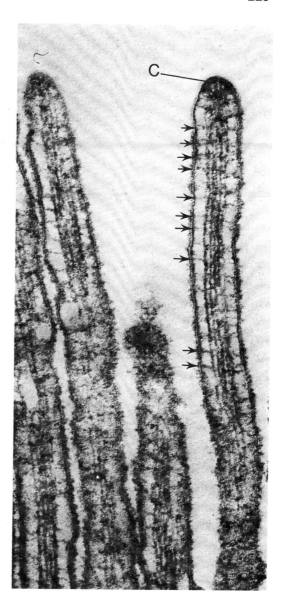

Figure 158. *A high-resolution micrograph of a micro-villus from a preparation similar to that shown in Figure 157. Here, in addition to the longitudinal bundles of actin filaments, cross-bridging elements with a regular periodicity are seen (arrows). The evidence suggests that these bridges, like the condensation (c) at the tip of the microvillus, consist of alpha actinin. As in the Z-lines of striated muscle, this rod-like protein is probably responsible for anchoring the actin filaments to the plasma membrane. It is of interest that the periodicity (33 nm) at which the alpha actinin cross-bridges are distributed corresponds with the estimated length of the tropomyosin filaments, which are known to be associated with (and probably lie alongside) the actin filaments. Magnification ×102 000.*

Courtesy of M. S. Mooseker and L. G. Tilney, Department of Anatomy, Harvard Medical School, Boston, and Department of Biology, University of Pennsylvania.

The brush border contractile system

The best documented actin-related contractile system in non-muscle cells is the arrangement found within the intestinal brush border. As shown in Figures 157 and 158, the cores of the microvilli that form this border consist of bundles of 5 to 7 nm filaments. The filaments contain actin and run the length of the microvillus from the tip (where they are attached by a condensation of alpha actinin — see below) to the base, where they extend into the apical cytoplasm. Within the apical plasm

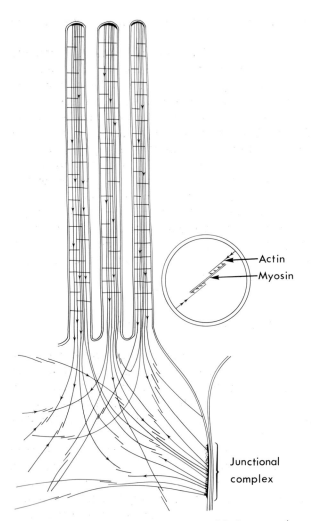

Figure 159. *A diagram summarizing some of the known and some of the postulated features of the filamentous arrangement in the brush border of the epithelium lining the intestine.*

Divalent cations (Ca^{++} and Mg^{++}) and ATP, when added to isolated brush borders, cause a contraction. In filamentous complexes isolated from these borders, similar treatment causes the actin bundles to plunge into and through the terminal web. It has thus been proposed that contractile force is generated in the brush border by the indirect association (see inset) of the actin filaments that emerge from the bundles in neighbouring microvilli with each other and with the actin and 10 nm intermediate filaments that arise from the junctional complexes at the lateral boundary. The polarity of the actin filaments is indicated by the arrow heads. The intermediary component with which these filaments interact is presumed to be myosin.

With the participating components attached to sites within the microvillus and within the lateral junctional complexes, contraction can be expected to cause a shortening and bending of the microvilli.

Drawing based on accounts by Mooseker and Tilney, and Roedwald, Newman and Karnovsky.

the actin filaments become distributed amongst the filamentous components of the 'terminal web', a mixed filamentous array that, in addition to 'thin' actin components, probably includes 'thick' myosin filaments. Within the terminal web there are also 10 nm, intermediate filaments derived from the well-developed junctional complexes situated on the lateral plasma membrane.

In the intact cell, contraction in this filamentous arrangement probably causes the microvilli to bend and shorten. However, this movement is not easily documented in the living mucosa (because the microvilli are at the limit of light microscope resolution) and direct observations have therefore been limited to those made on fragmented systems fixed immediately after contraction has been induced (by the addition of divalent cations and ATP). In these 'contracted' preparations the actin bundles of the microvillous core are seen to have plunged into and through the filaments of the terminal web, and it thus seems clear that force generation arises from an interaction within the web complex. Heavy meromyosin binding shows that the polarity of the actin filaments that emerge from the microvillous core is directed away from their site of attachment and towards the force-generating interactions (see Figure 159) within the terminal web. Their arrangement is thus essentially the same as that found in the striated muscle sarcomere.

Microfilament networks at the cell periphery

At the cell periphery, actin is distributed within the network of 5 to 7 nm microfilaments that lies beneath the plasma membrane of most free cells (Figure 160). It is probable that the mobility of some integral components of the plasma membrane is controlled by this network, and it is also likely that the fluidity of the peripheral cytoplasm is determined by its degree of polymerization.

The role of peripheral microfilaments in controlling the mobility of plasma membrane components has been most clearly indicated in studies in which local movements of the cell membrane have been shown to be inhibited by cytochalasin B, a fungal metabolite that is believed to interfere directly with actin-containing microfilaments. In macrophages, for example, where cytochalasin B inhibits phagocytosis, it is thought to act by disturbing the contractile mechanism responsible for drawing the invaginating plasma membrane inwards.

The peripheral microfilamentous network almost certainly plays an important role in cell locomotion (when a free cell moves over a solid substratum – Figure 161), but the details of the way in which it participates in the locomotory process are likely to be very complex. It is, for example, probable that the increased fluidity of the peripheral cytoplasm, required for the extension of the leading edge of a moving cell, depends upon the lability of the microfilamentous network; indeed immunocytochemical studies showing a localized rearrangement of the actin filaments in that area support this idea (Figure 162). At the same time, however, it seems that the grip of the surface below the leading edge, which is exerted upon the substratum, requires local stiffening, and electron microscope studies suggest this is provided by the formation of narrow focal concentrations of microfilaments.

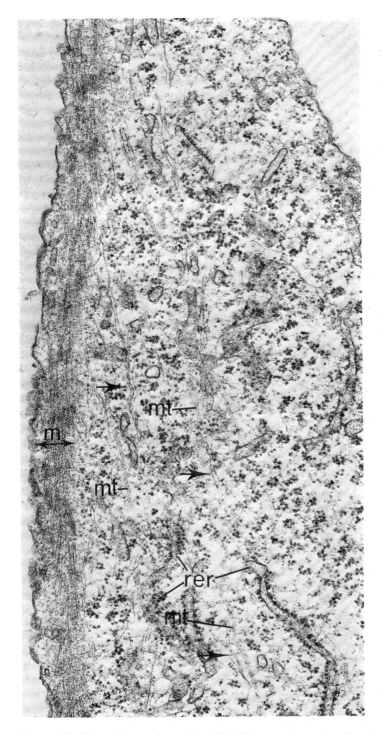

Figure 160. *The periphery of a cultured fibroblast seen in section. Running parallel to the plasma membrane is a dense feltwork of microfilaments (m). Deeper in the cytoplasm are occasional thicker (10 nm) intermediate filaments (arrows), several short profiles of microtubules (mt), and scattered elements of rough endoplasmic reticulum (rer). Magnification ×24 000.*

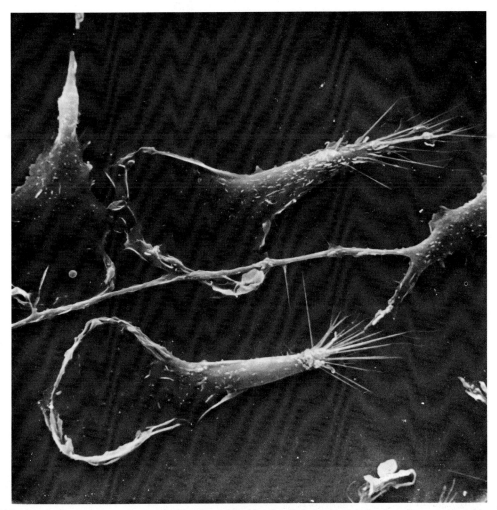

Figure 161. *A scanning electron micrograph showing white blood cells (monocytes) fixed as they were moving over a flat 'non-biological' surface. The leading edge of each cell is spread and flattened, while more posteriorly, in the region containing the nucleus, the cell 'body' narrows and becomes spindle-shaped.*

It seems that as the body of the cell moves forward, its lower surface remains adherent to the substratum and the cell membrane is pulled into an array of fine radiating strands (often called 'retraction fibres'). Although this prolonged adhesion probably occurs only when cells move over an artificial surface, the distribution of the attenuated strands suggests that the contact the lower surface of a moving cell makes with the substratum is restricted to small, localized foci.

As in the fibroblast, the microfilamentous components responsible for movement in leucocytes include actin and its related proteins. There is also evidence, however, that in leucocyte movement microtubules are also important and it seems likely that there may be important differences in the way in which the cytoskeletal and contractile components participate in these different groups of cells. In passing, it is also worth mentioning that the speed of moving leucocytes (about 10 to 12 μm/minute) is about ten times faster than that of the cultured fibroblast. Magnification $\times 1500$.

Courtesy of T. Allen, Paterson Laboratories, Christie Hospital and Holt Radium Institute, Manchester.

To these localized involvements there must be added the possibility that there is also a positive force for forward movement that also relies upon force generation by microfilamentous elements.

Currently, detailed studies of these membrane-related processes are hindered because actin-containing microfilaments at the cell periphery

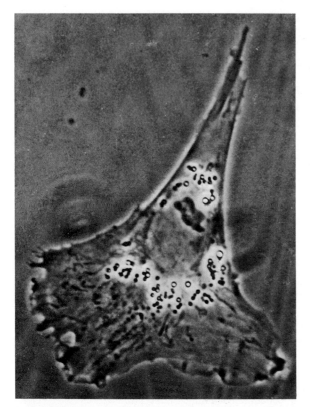

(a)

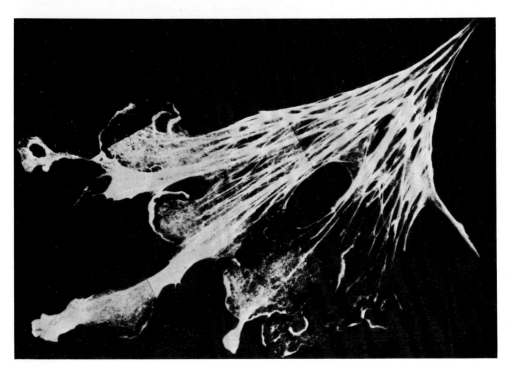

(b)

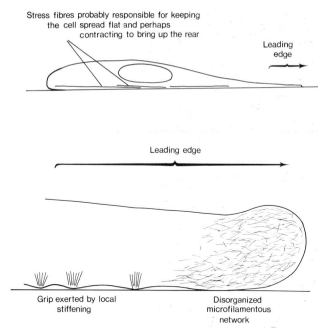

Stress fibres probably responsible for keeping the cell spread flat and perhaps contracting to bring up the rear

Leading edge

Leading edge

Grip exerted by local stiffening

Disorganized microfilamentous network

(c)

Figure 162. *Locomotion in actively mobile cells.*

The major defence systems of the body rely to a large extent upon single, free cells such as blood leucocytes and macrophages that are able to migrate actively through the tissues. Movement in these cells can be followed in culture, and it is currently being studied in considerable detail.

To a lesser extent other cell types such as the fibroblasts of the connective tissues also migrate. The movement of these cells is less easily demonstrated in vivo, but because their form and behaviour in tissue culture makes them easily approached by the microscopical and biochemical techniques presently available they, too, are being widely used for studies of the locomotory process.

(a) (Opposite.) This phase contrast micrograph shows a fibroblast in culture. The cell is moving towards the bottom left-hand side of the field and is about to turn towards the right-hand corner. Note the ruffles at the leading edges and the narrowing of the cell at the rear. Within the cell the filamentous mitochondria, highly refractile lipid droplets, and the ovoid nucleus are all well shown.

Courtesy of M. Abercrombie, Strangeways Laboratory, Cambridge.

(b) (Opposite.) The distribution of actin within a moving cell, shown by immunofluorescence. The cell is moving to the lower left. Below the leading edges the actin is distributed as a diffuse network, whereas, more posteriorly, it is organized into coarse stress fibres that run in the long axis of the cell.

Courtesy of R. Pollack, Department of Microbiology, State University of New York at Stonybrook.

(c) (Above.) A diagram indicating the distribution of some of the actin-containing components in an actively moving fibroblast.

are much less stable and more difficult to preserve than those elsewhere in the cell. The instability of these filaments suggests that they may be organized differently from the microfilaments of the more highly ordered contractile systems, and it may reflect the fact that they are indeed transitory elements that readily move between a filamentous phase and a soluble phase. The polymerization and depolymerization of filamentous F-actin from a reservoir of soluble G-actin is regarded as providing the most likely molecular basis for this interchange. Similar transitions between an ordered polymerized state and a pool of soluble subunits are believed to exist within the cytoplasmic matrix for other, related skeletal components (e.g. myosin and tubulin).

Components related to actin

Immunocytochemical studies have shown that in many non-muscle cells the actin-containing filaments may also contain associated 'thin' myofilament components such as tropomyosin and alpha actinin. In some instances, as in the stress fibres of cultured fibroblasts, these two proteins show a regular, alternating periodicity with each other.

Macrophage
Platelet

In cells such as the macrophage and the blood platelet, myosin is known to be a major component of the cytoplasmic matrix, while immunofluorescence indicates that this protein also occurs in the mitotic spindle (see page 238), below the equatorial furrow at cytokinesis, and in the stress fibres of cells in culture. It seems reasonable to assume that in these circumstances movement depends upon a sliding filament mechanism similar to that occurring in muscle.

Elsewhere, however, a myosin component has not yet been identified with certainty and while it is possible that it is present in very low concentrations relative to actin, the alternative possibility, that a filamentous protein other than myosin may be capable of associating with actin and generating contractile force, cannot be excluded. Filamin is currently the best candidate for this role.

Insertion into the plasma membrane

Intermediate (10 nm) filaments are a common and often major feature of many non-muscle cells. Amongst the most abundant are the tonofilaments that occur as coarse bundles inserting into desmosomal plaques on the plasma membrane in epithelial cells (Figure 63). The chemical nature of these filaments has not yet been established, but it is probable that they are similar to the other, less abundant 10 nm filaments that are found within the body of most cell types (often in the company of microtubules). Immunofluorescence suggests that these 10 nm filaments are chemically similar to the desmin-containing, intermediate filaments of smooth muscle. The relationships between the 10 nm filaments of non-muscle cells and force-generating components like actin and myosin is unclear, but at the present time their function is believed to be primarily concerned with anchorage and attachment to the plasma membrane.

In cell locomotion and other surface-related events, such as phagocytosis, the filaments of the contractile mechanisms also need to be attached to the plasma membrane. It is possible that in some circumstances the actin filaments may attach to integral membrane proteins directly, but in others it seems clear that, as in muscle cells, there are intermediary anchoring components.

In brush border microvilli there is good reason to believe that these intermediary components can be identified directly. Thus, as shown in Figure 158, actin filaments within the core of the microvillus insert at the tip into cap-like condensations and along the sides via thin, rod-like cross-bridges. Immunofluorescence shows that both the condensations at the tip and the cross-bridges contain alpha actinin. Indeed the size and form of the latter (2×30 nm), as seen in the electron microscope, conform to the estimated size of this molecule.

In erythrocytes yet another filamentous protein called 'spectrin' (mol. wt. about 220 000) has been identified. This component, although a major peripheral membrane protein, has thus far only been found in erythrocytes. It does, however, have many chemical and physical features in common with filamin, and of special interest in the present context is the fact that it, too, is believed to be able to interact directly with integral components of the plasma membrane.

Within nerve cells there are abundant 10 nm filaments, known as 'neurofilaments'. These filaments are distributed linearly along nerve fibres and, as shown in Figure 109, in some ganglion cells they may also be distributed throughout the nerve cell body. The relationship between these 10 nm filaments and the prominent 'neurofibrils' identified by light microscopy is unclear. Unfortunately, although it has been established that neurofilaments are chemically different from tonofilaments, little is known of either their chemical nature or their function.

MICROTUBULES

Microtubules are a class of long, unbranched, tubule-like elements which are found in all types of cell. They may exist in highly ordered arrangements, as when they form the central core (or 'axoneme') of cilia, or they may be distributed, apparently without preferred orientation, amongst the other organelles in the cytoplasmic matrix. Formed microtubules, as seen in the electron microscope in aldehyde-fixed preparations,

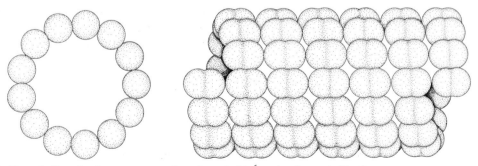

Figure 163. *The tubulin subunit arrangement of a microtubule, seen in plan and side view.*

have an outer diameter of 25 nm, with a central, apparently hollow core (4 nm in diameter). Their walls are composed of subunits of the protein, *Subunit* tubulin (Figure 163). Each tubulin subunit has a molecular weight of *Protomer* about 120 000 and is composed of two chemically related protomer molecules (alpha and beta tubulins). The tubulin subunits are arranged linearly to form thirteen, 5 nm diameter 'protofilaments', spiralling with a regular periodicity of 80 nm as a three start, left-handed helix. Like actin filaments, microtubules probably have an intrinsic polarity, a property that determines the direction in which they become displaced when they slide and move relative to one another.

Microtubules in the axoneme of the cilium

Microtubules are the major skeletal and force-generating elements of cilia (Figure 164). In motile cilia the microtubules within the axoneme are arranged in a characteristic formation — nine peripheral doublets

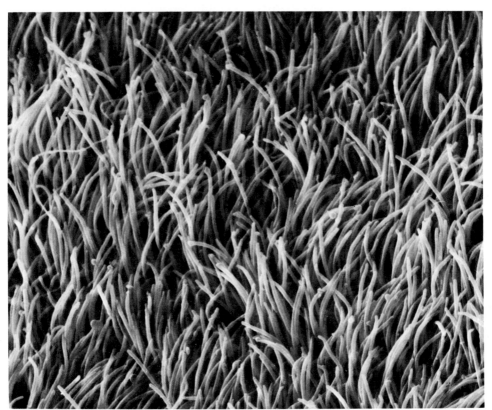

Figure 164. *A scanning electron micrograph of cilia on the apical surface of ependymal cells lining a ventricle of the brain. In life, this ciliated carpet, like that lining the mucosal surfaces in the respiratory and reproductive systems, probably beats with a synchronized metachronal rhythm (each individual cilium moving slightly ahead of the one behind it) to provide a regular, directed wave of movement across its surface. Magnification ×4500.*
Courtesy of P. W. Coates, Department of Biological Structure, University of Washington, Seattle.

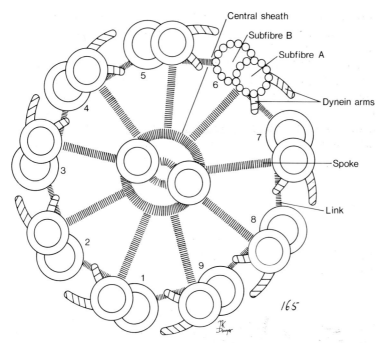

Figure 165. *A cross-sectional view of the axoneme to show the interrelationships of the major components.*

around a central pair. The doublet arrangement of two longitudinally aligned microtubular elements appears to be unique to cilia (and cilia-like structures such as the tail of the spermatozoon) and has been examined in considerable detail (Figures 165 and 166).

In each of the peripheral nine doublets there is one microtubule composed of 13 protofilaments (subfibre A) associated with an incomplete microtubule of 11 protofilaments (subfibre B). Extending from subfibre A towards the next adjacent subfibre B are pairs of arm-like processes composed of the protein, dynein, a high molecular weight ATPase. The dynein arms are regularly arranged along the length of the A-subfibres. Other arm-like processes ('spokes' and 'links' – see Figure 166) in regular array connect the subfibre with a 'central sheath', which surrounds the central pair of microtubules. They also connect their neighbouring peripheral doublets.

In a cross-sectional view of the axoneme (Figure 165), it is apparent that the alignment of the central pair of microtubules and the linkage between the peripheral doublets designated '5' and '6' bestows a bilateral symmetry upon the structure. It is along the axis of this symmetrical arrangement that the beating cilium bends. Although non-motile cilia often lack the central pair of microtubules, the available evidence suggests that these elements are concerned with the alignment of movement rather than its propagation. Force generation in cilia is more likely

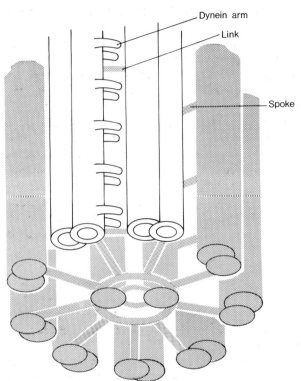

Figure 166. *A three-dimensional view of the axoneme, as seen from the base of the cilium.*
    The dynein arms of subfibre A have a periodicity of 22.5 nm. Sliding is believed to depend upon a cyclic attachment of the dynein arms on the A-subfibre of one doublet to the B-subfibre of the adjacent doublet, so that the arms 'walk' along it.
    The links that attach each A-subfibre to the B-fibre of the adjacent doublet have a periodicity of 86.5 nm, and they are believed to be made of an elastic material (called 'nexin'). They are not thought to detach when the doublets slide.
    The arrangement of the spokes is more complex, since they appear to be arranged in groups of three to form a double spiral around the central microtubule pair. When the cilium bends, they probably detach and then reattach.

to depend upon the sliding of one peripheral microtubule doublet in respect to its neighbour. Like the actin–myosin interaction, movement in cilia is known to require ATP and calcium ions, and the most widely accepted model of axonemal movement proposes that the peripheral doublets derive energy from the activity of their dynein ATPases and slide relative to each other, causing linear shear forces that induce the cilium to bend. Relaxation is presumably passive and may, perhaps, depend upon the elasticity of other axonemal components (such as the links) or the enveloping plasma membrane. The precise role of calcium in the control of cilial movement is unclear, but it would not be surprising if, as in muscle cells, it was found to act as an initiating signal.

Polymerization of microtubules and the effects of colchicine

In addition to the stable and highly regimented arrays of microtubules in cilia, microtubular elements are a prominent feature of mitotic spindles and are also found in large numbers within the axons and dendrites of neurones in the central nervous system. In rather fewer numbers they also occur as a ubiquitous component beneath the plasma membrane and in the juxtanuclear Golgi area of many cell types. In these situations microtubules are less stable than those of the ciliary axonemes, and their length (that is, their degree of polymerization) is directly dependent upon the equilibrium of their tubulin subunits with the free pool of unpolymerized tubulin within the cytoplasmic matrix. Changes in the intracellular environment (such as changes in local ion concentration) are probably able to move the balance of the equilibrium either towards or away from polymerization, with the result that microtubules are either induced to appear and increase in length or, alternatively, to disperse. The dynamic state of interchange between formed microtubules and the pool of tubulin subunits is perhaps associated with the rapid and often local changes in cytoplasmic fluidity which occur in mobile cells. It also has a primary role to play in altering cell shape. As described below, the dramatic appearance of spindle elements before cell division also draws upon the cytoplasmic tubulin pool.

Colchicine, an alkyloid from the autumn crocus, *Colchicum autumnale*, will interfere with the tubulin equilibrium because it binds to the free tubulin subunits of the cytoplasmic pool. The change in equilibrium causes the microtubules to break down, and the events with which they are concerned may be arrested. Thus, for example, colchicine acts as a 'spindle poison' since it prevents the orderly handling of chromosomes at mitosis. Because this alkyloid will also inhibit other cellular activities, such as the stimulation of cell division in cultured cells and the secretion of serum protein by liver parenchyma cells, it has been suggested that these processes are also, in some manner, dependent upon the presence of formed microtubules.

The axonemes of cilia are unaffected by colchicine, indicating that these microtubular components are not influenced by changes in the cytoplasmic pool of free tubulin. When cilia grow, they must, of course, depend upon tubulin synthesized in the cytoplasm. However, it is known that when tubulin subunits are added to an axoneme, they are added distally, at the tip of the cilium, not at the base, in the cytoplasm.

Microtubules and mitosis

At cell division, the orderly segregation of the divided chromosomes (the 'chromatids') between the daughter cells is accomplished by virtue of

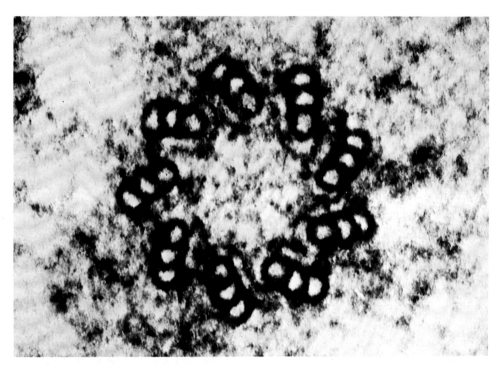

Figure 167. *An electron micrograph showing the details of a centriole in cross section. The nine-fold symmetry of the microtubule triplets is clearly shown, and there also appears to be an inner, central arrangement with the same symmetry. The significance of these elements is unknown. Magnification ×300 000.*
  Courtesy of E. de Harven, Sloan-Kettering Institute for Cancer Research, New York.

their arrangement on the mitotic spindle. Since the spindle has a micro-tubular skeleton, studies on the function of microtubules have, from the beginning, paid a good deal of attention to this structure during mitosis.

Centrioles and microtubule-organizing centres

In early mitosis the position and form of the mitotic spindle is dictated by two microtubule-organizing centres, the 'centrioles'. The nature and origin of these organelles is obscure, although their structure, which has a nine-fold symmetry, is highly organized and very characteristic (Figure 167). Their structure is closely similar to that of the 'basal bodies', which

Figure 168. *(Opposite.) The apical boundary of the ciliated cells lining the oviduct. The cilia beat towards the uterus, and, together with the muscular activity of the oviduct wall, they aid the passage of the ovum.*
  Where the axoneme (ax) of each cilium arises from the cell body, it is associated with a basal body (bb). The organization of these structures is essentially the same as that of centrioles, except they are often associated, as here, with prominent, striated rootlets. The nature of the fibrous elements that make up the rootlet bundle (rb) is not known. Presumably they serve only to anchor and stabilize the base of the axoneme. Magnification ×25 000.
  Courtesy of N. Bjorkman, Department of Anatomy, Royal Veterinary and Agricultural University, Copenhagen.

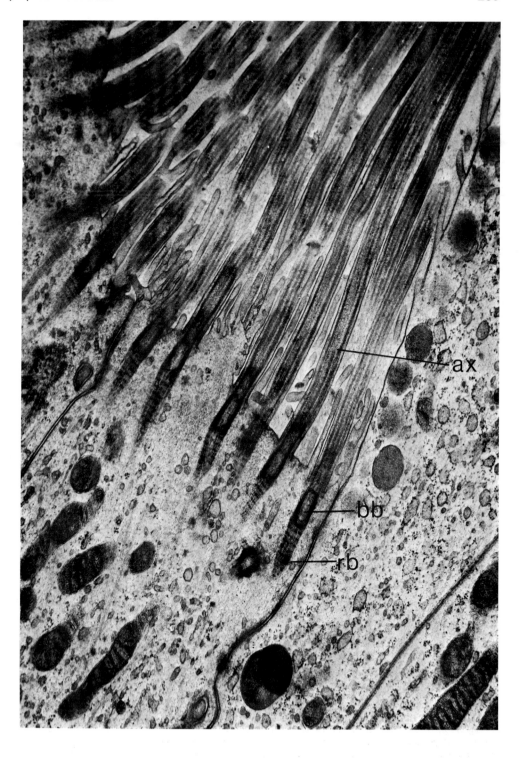

Figure 168

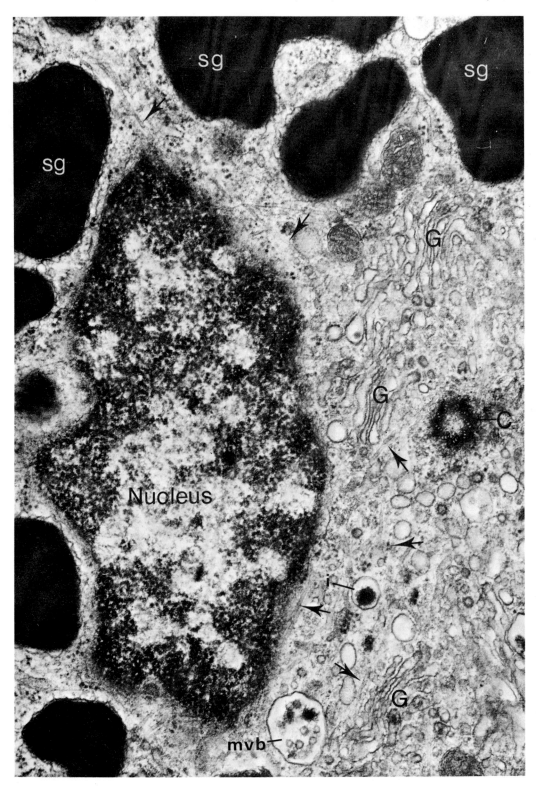

Figure 169

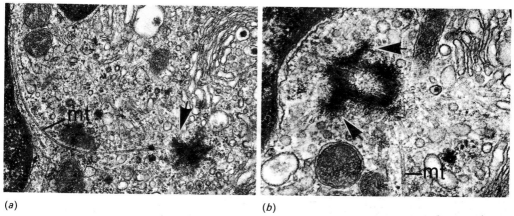

(a)   (b)

Figure 170. *Further details of the mast cell cytocentrum. In both micrographs the centrioles are sectioned along their length and the nine-fold symmetry of their microtubular components is thus not displayed. However, their lateral condensations, known as 'satellite bodies', are apparent (arrows). It is from these condensations that the associated microtubular elements (mt), which radiate into the surrounding cytoplasmic matrix, arise. Magnifications: (a) ×27 800; (b) ×48 000.*
    Courtesy of D. Lagunoff and Y. Chi, Department of Pathology, University of Washington, Seattle.

occur at the base of cilia and to which the centrioles give rise (Figure 168). Typically, centrioles occur in the 'cytocentrum', a non-labile area of the cytoplasmic matrix which includes the Golgi area (Figures 169 and 170). They usually occur in pairs situated at right angles to each other and, like the basal bodies, they seem to divide autonomously. Centrioles and basal bodies attach more or less directly to microtubules and they are a preferred site within the cell for microtubule polymerization. Kinetochores, the specialized attachment regions on the centromeric region on chromosomes, also act as microtubule-organizing centres. As yet, however, nothing is known of the way in which any of these structures cause microtubules to form.

At the beginning of mitosis (see Figures 171 to 174) centrioles divide, move into position at the prospective poles of the spindle, and microtubule formation begins. Then, following the breakdown of the nuclear membrane, they and the kinetochores of the condensed chromosomes (which are now exposed to the pool of cytoplasmic tubulin) induce the further formation of very large numbers of microtubules. These microtubule elements extend both from centriole to centriole (pole to pole) and from kinetochore (50 to 60 per kinetochore) to centriole. It is likely that the demand made upon the cell's resources of tubulin at

Figure 169. *(Opposite.) The Golgi area in a mast cell. In addition to the large, electron-opaque secretory granules (sg), the juxtanuclear cytoplasm contains the diverse population of elements characteristic of the area known as the cytocentrum. Providing a focus within it are the two centrioles (C) (only one is shown here – in cross section), and radiating outwards from them are numerous microtubules (arrows). The complexity of the area is compounded by various Golgi elements (G), which include flattened, parallel cisternae, small vesicles, an immature secretory granule (i), and, near the lower margin, an element of unknown function, a multivesicular body (mvb). Other organelles, such as the rough endoplasmic reticulum, mitochondria, and mature secretory granules, are normally excluded from the inner centriolar region, but, as shown here, they may encroach at the periphery. Magnification ×46 700.*
    Courtesy of D. Lagunoff and Y. Chi, Department of Pathology, University of Washington, Seattle.

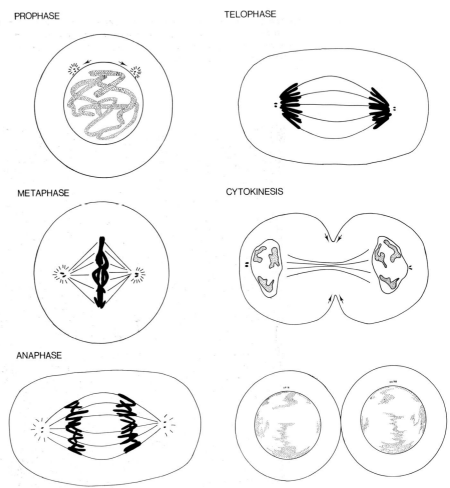

PROPHASE

TELOPHASE

METAPHASE

CYTOKINESIS

ANAPHASE

Figure 171. *Mitosis in outline. See Figures 172 to 178 for the formation and distribution of the spindle components during the different phases of mitosis.*

Cytoskeleton

this time draws heavily upon the other cytoskeletal microtubular elements, and it is probably for this reason that, in monolayer culture, dividing cells usually lose their flattened shape and assume a spherical 'rounded-up' form.

Clearly, the arrangement of the microtubules in the mitotic spindle ensures that when the chromatids move, they move to opposite ends of the cell (see Figures 175 to 178). However, although it is known that the part played by the chromatids themselves is passive, it is not clear if the microtubules actively participate in generating force for their movement. Several hypotheses ascribing a sliding tubule or polymerization–depolymerization mechanism have been put forward to account for the roles that microtubules may play, but at the present time there is no definitive evidence in favour of any of them.

Figures 172 to 174 *show the events which occur during mitosis in the cell line Pt-K. This cell line has been widely used for the study of cell division because its cells have the unusual characteristic of remaining flat when they divide. Phase contrast microscopy can thus be used to follow the division process in detail and individual cells can be selected for further examination in the electron microscope.*

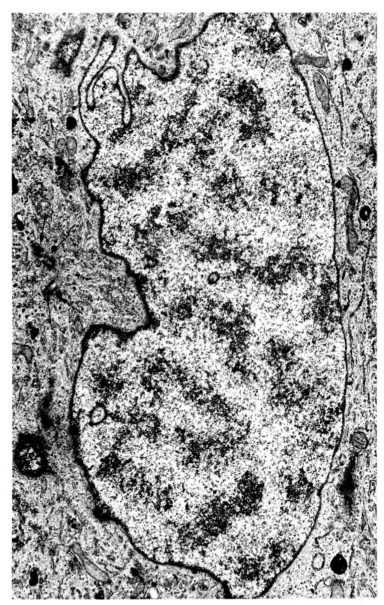

Figure 172. *In early prophase, while the nuclear membrane remains intact, the chromosomes condense so that they appear in section as opaque, aggregated masses. On the surfaces of these chromosomes the kinetochores (not apparent here) are represented only as small superficial patches of filaments. Note the paired centrioles that have divided but not yet migrated (arrows). Magnification ×10 000.*
 Courtesy of B. R. Brinkley and E. Stubblefield, Department of Cell Biology, Baylor, College of Medicine, Texas.

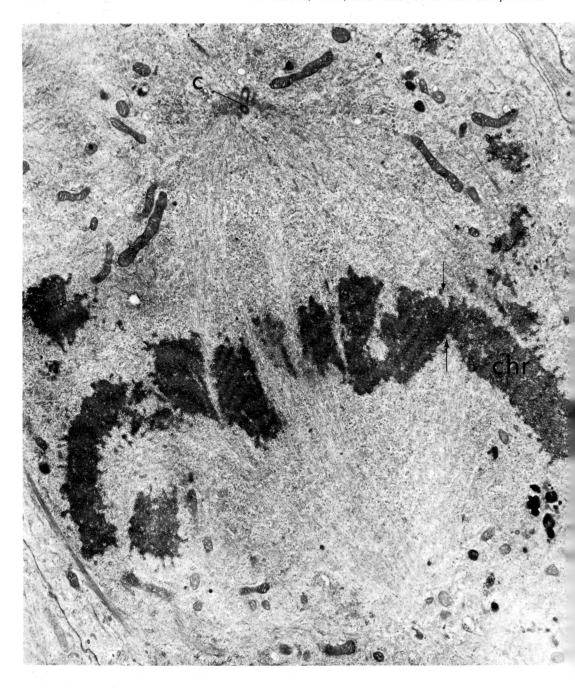

Figure 173. *As the spindle takes shape the chromosomes become aligned about its equator. This rearrangement marks the onset of metaphase and is called 'metakinesis'.*

In this micrograph note the microtubules of the spindle and the single pair of centrioles (c) (the other pair are not included in this section plane). One chromosome (chr) is sectioned almost longitudinally and the narrowed centromere region (where the kinetochores are situated) is clearly seen (arrows). Magnification ×10 200.

Courtesy of B. R. Brinkley and J. Cartwright, Department of Cell Biology, Baylor College of Medicine, Texas.

Figure 174. *Microtubules inserting into the kinetochores of two chromatids. The chromatids are cut in cross section and consist of a mass of condensed chromatin fibres. The arrows indicate the line along which they would have separated. Although the chromatid centromeres are undistinguishable here, their position is indicated by the plaque-like condensations of the kinetochores (K). The parallel, ribbon-like groups of microtubules are seen inserting into their respective kinetochores. Magnification ×50 000.*

Courtesy of J. R. McIntosh, Department of Molecular, Cellular and Developmental Biology, University of Colorado.

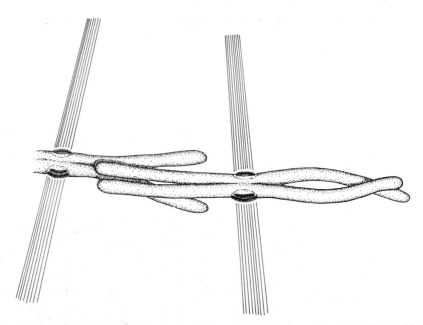

Figure 175. *Each kinetochore is situated in a notch-like depression in its chromatid and consists of two parallel brush-like filaments (200 to 500 nm long). It is probable that the microtubules that end within the immediate vicinity of these filaments are responsible for directing the separated chromatids towards their respective spindle poles, although it is not known how these microtubules accommodate for this movement. At the present time the best evidence favours the idea that as the chromatids move towards the poles, their associated microtubules gradually become shorter — probably by depolymerizing at the ends nearest the centrioles.*

To complicate the situation, it seems that there are at least two kinds of movement within the spindle with which microtubules might be involved. One is concerned with actually moving the chromatids towards their respective poles, while the other concerns the elongation of the spindle itself (at anaphase it more than doubles in length). The movement of the chromatids is obviously most easily related to the microtubules that insert into the kinetochores, but since it has been shown that actin and probably myosin also occur within their immediate vicinity, it is still possible that these microtubules are concerned only with guiding rather than pulling the moving chromatids.

Other functions

Although microtubules are also prominent features in cells where they are clearly not related to the movement of cilia or the mitotic spindle, their role in these situations is not entirely clear. It is certain, however, that in many cells the microtubules are important skeletal elements that determine cell shape, and this is particularly true where microtubule bundles are arranged at the cell periphery (Figure 179). In this context those of platelets that provide the 'marginal bundle' and maintain the discoid shape of these cell fragments are a good example.

Figures 176 to 178 *show the later stages of mitosis as seen in cultured fibroblasts.*

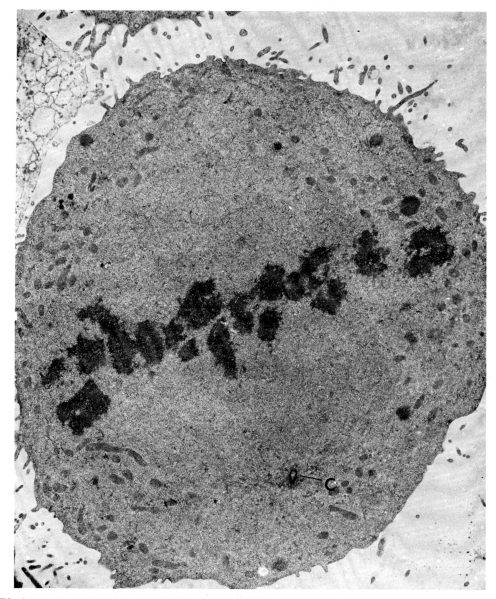

Figure 176. *At metaphase the cell has a spherical shape and there are many surface microvilli (see also Figure 76). The condensed chromosomes (paired chromatids) are arranged at the spindle equator. Individual microtubules are below the resolution of this micrograph; one of the polar centrioles can, however, be identified (c). Magnification ×12 500.*

Courtesy of B. R. Brinkley and P. N. Rao, Department of Cell Biology, Baylor College of Medicine, Texas.

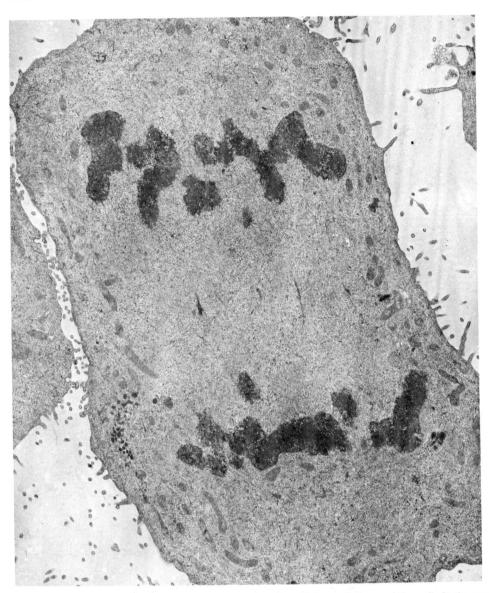

Figure 177. *In anaphase the spindle elongates and so does the overall shape of the cell. At the same time the individual chromatids separate and give rise to two groups, which then move apart along the elongating spindle towards the poles. Magnification ×12500.*

In telophase, as they approach the centrioles, the chromatids become encircled by vesicular elements and these then fuse to form the nuclear envelope. Soon afterwards the condensed chromatids disperse and their kinetochores disappear.

Courtesy of B. R. Brinkley and P. N. Rao, Department of Cell Biology, Baylor College of Medicine, Texas.

Elsewhere, however, while it is probable that microtubules may also be concerned with intracellular transport, it has not been possible to define their role. Even in tissues like the nervous system, where directed cytoplasmic transport (axoplasmic flow) has been clearly demonstrated, and where neurotubules (which are microtubules) are a consistent and

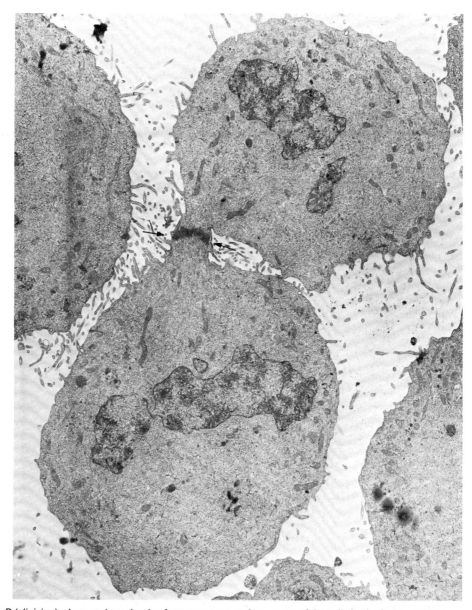

Figure 178. *In D (division) phase an invaginating furrow appears at the equator of the cell. As the furrow progresses inwards (cytokinesis) it produces a dumb-bell shaped outline in which the emerging daughter cells remain attached by a narrow waist. This narrowed region is often called the 'mid body' and contains the remnants of the continuous spindle microtubules embedded within an electron-opaque condensate (arrows). The chemical nature of the condensed material within the mid body and its role, if any, in the final severance of the bridge connecting the daughter cells is unknown.*

Within the nuclei at this time the chromatin has assumed its characteristic interphase distribution and the nucleoli have reappeared. Magnification ×9000.

Courtesy of B. R. Brinkley and J. P. Chang, Department of Cell Biology, Baylor College of Medicine, Texas.

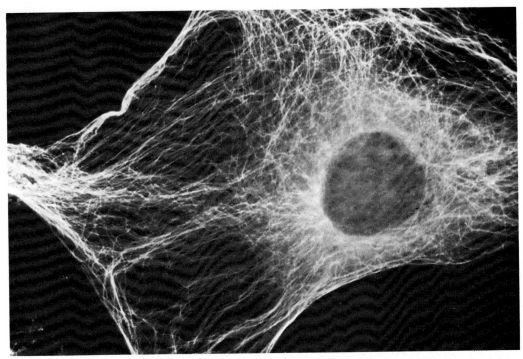

Figure 179. *The distribution of microtubules in a cell in culture demonstrated immunocytochemically using an antibody directed against tubulin. The immunofluorescence shows a dense population of microtubules concentrated in the perinuclear area. These components clearly have a very different distribution from that of actin filaments (see Figure 155). Individual microtubules do not branch and with a diameter of 25 nm they are well below the resolution of the light microscope. The network seen here thus probably represents interconnecting bundles of microtubules.*

Courtesy of K. Weber, Department of Biochemistry, Max-Planck Institut für Biophysikalische Chemie, Goettingen, Germany.

very prominent feature (tubulin may account for as much as 15 per cent of the total cell protein), there is no definitive information on their role in the transport process.

An additional problem has been the difficulty of showing whether or not microtubules interact directly with membranes. Thus, although it is widely accepted that microtubules are involved in many membrane-related processes like phagocytosis, the main weight of the proof that they do so remains circumstantial because it relies almost entirely on the sensitivity of these processes to colchicine treatment.

GLOSSARY

Definitions of the italicized marginal entries.

Acidic stains. Histological stains that are negatively charged at the pH used for staining are said to be 'acidic'. Eosin is an acidic stain.

Acinus. Group of secretory cells emptying into a common duct (From the Latin, 'a cluster of berries').

Adenosine triphosphate (ATP). Molecule containing adenine, ribose and three phosphate groups. Its singular importance as an energy store depends upon its two high-energy phosphate bonds. On hydrolysis, these bonds release a large amount of free energy.

Amino acids. The building blocks of proteins. All of them contain a carboxyl group and an amino group but they differ in the chemical nature of their side (R) groups. Classified according to the polarity of their side groups, the non-polar (hydrophobic) class includes leucine, proline and valine. The polar class contains glycine, serine, tyrosine, arginine and glutamic acid.

Analgesic. An agent producing pain relief.

Antibodies. Antigens can induce 'B' lymphocytes to transform into cells (plasma cells) that secrete immunoglobulins. Immunoglobulins are a specific class of glycoprotein molecule that can identify and bind to an antigen. They circulate in the blood and other body fluids. Blood (with blood cells removed) which contains antibodies raised in response to a specific antigen constitutes an 'antiserum'.

Antigen. Any molecular feature that an immune system regards as foreign. Antigens stimulate an immune response and this response often includes the production of an antibody specific for the antigen.

Apical membrane. Epithelial cells that line ducts (and tracts) often have a pyramidal shape. The membrane at their blunt apex (the apical membrane) normally borders on the duct lumen.

Basic stains. Histological stains that are positively charged at the pH used for staining are said to be 'basic'. In general, these stains (e.g. methylene blue) stain components such as nucleic acids which are negatively charged at this pH. These *components* are said to be 'basophilic'.

Brownian motion. Random motion of small particles in suspension which results from their being bombarded by molecules in the surrounding medium.

C terminus. See *Peptide bond*.

Carcinogen. Exogenous chemical agent that causes cancer.

Cathepsins. A group of acid proteases (i.e. those proteases with an optimum activity at acid pH) present in lysosomes.

Cell-free systems. Broken cell preparations, often composed of a single subcellular fraction, in which limited aspects of cellular metabolism can be followed in isolation.

Cell line. Cells surviving repeated subculture. The cells of cell lines always show altered characteristics.

Centromere. That part of the divided chromosome containing little or no 'informational' DNA. At mitosis it is concerned with attachment (via the kinetochores) to the mitotic spindle.

Cholinergic nerves. Neurones that release the neurotransmitter acetylcholine. Good examples are the motor neurones that innervate skeletal muscle.

Chondrocyte/Chondroblast. Chondrocyte, a cell living within the cartilage matrix; chondroblast, a cartilage matrix-forming cell.

Chromatids. At the beginning of mitosis each condensed chromosome is seen to be composed of duplicate strands joined at their centromeres. Each strand is a chromatid.

Chromosomes. Thread-like structures containing the DNA in which most of the cell's hereditary information is encoded. In eukaryotes the full complement of this information (the genome) is divided between several chromosomes. In prokaryotes it is contained within a single, circular structure.

Chylomicron. Blood-borne particle consisting of lipid, protein and carbohydrate. Chylomicrons carry newly synthesized triglyceride from the intestine to the liver.

Cisterna. A closed, membrane-limited system, sometimes tubular, sometimes sac-like in form.

Codon. Messenger RNA consists of a chain of nucleotide bases. During translation each consecutive group of three bases is responsible for the selection and incorporation of one amino acid. Each group of three bases is a 'codon'.

Coenzymes. Low molecular weight compounds that bind to enzymes and permit them to react with their substrates.

Coenzyme A. Serves as a carrier for acyl groups in enzymic reactions in the tricarboxylic acid cycle. Acetyl-coenzyme A, which is formed during the enzymic oxidation of pyruvate or fatty acids, contains a high-energy bond.

Complement. A complex of at least nine components present in serum. It is frequently called upon to participate in the immune response and when activated it can cause death and phagocytosis of 'foreign' cellular components.

Connective tissues. Tissues (like cartilage and bone) that connect and maintain the structural integrity of the body. They are derived from embryonic mesenchyme and are typified by a large amount of extracellular material (the matrix within which the constituent cells reside).

Constitutive proteins. Proteins that in a given cell are always present in constant amounts. Most often used with reference to certain prokaryotic enzymes, where synthesis seems to be independent of growth conditions.

Covalent bond. Strong bonds (like peptide bonds) that depend upon two atoms sharing electrons.

Cyclic AMP. Adenosine monophosphate with a phosphate group bonded internally to form a cyclic molecule. An important 'second messenger' formed from ATP by the enzyme adenylate cyclase. Many hormones (e.g. glucagon, adrenocorticotrophin and gonadotrophins) stimulate their target cells by stimulating the adenylate cyclases within their plasma membranes.

Cycloheximide. Inhibits protein synthesis by preventing elongation of the growing peptide chain.

Cytokinesis. Follows mitosis, when the inward movement of the cell surface forms an invaginating furrow to divide the cell in two.

Cytoplasmic basophilia. In cells synthesizing large amounts of protein, the cytoplasm stains strongly with 'basic' stains. This 'cytoplasmic basophilia', as it is called, is due primarily to the RNA contained within the dense population of polyribosomes which characterizes the cytoplasm of these cells.

Cytoskeleton. Intracellular components which support and determine cell shape and form. The most important are the microtubule and microfilament networks.

Dalton. Unit of weight equal to the weight of a single hydrogen atom.

Diploid. Containing two complete sets of chromosomes.

Endocrine/Exocrine. Exocrine secretory cells release their secretory product into ducts lined with epithelia, while endocrine cells usually direct their released secretory product towards the bloodstream. Endocrine secretions always include hormones.

Endogenous. Of internal origin.

Epidermis. Outside layer of the skin consisting of multiple strata of epithelial cells. As the more superficial cells are sloughed away they are continuously replaced from below by the division of stem cells within the more basal (germinative) layers.

Epithelium. Lining and glandular tissues derived from embryonic endoderm and ectoderm. Consists of adherent cells with little intercellular matrix.

Erythroblast. Developing cells that eventually give rise to mature enucleate erythrocytes.

Escherichia coli (E. coli). A non-pathogenic bacterium found in the intestines of man. The most intensively studied and best-understood organism in molecular biology.

Eukaryotes. Organisms and cells in which the genome is contained within a special compartment, the nucleus. A highly organized cytoplasmic compartment containing organelles is also a characteristic eukaryotic feature.

Exocrine. See *Endocrine*.

Fibroblasts. Race of cells inhabiting the connective tissues of the body. Responsible for synthesizing and secreting non-cellular components such as collagen. They survive well and actively proliferate in tissue culture.

Filamentous proteins. Proteins in which the polypeptide chains are arranged in parallel along a single axis to form fibres or sheets (e.g. collagen).

Gene/Structural gene. Part of the hereditary information embodied in the DNA of a chromosome. A structural gene is that part of the DNA transcribed to provide an mRNA for translation in the cytoplasm.

Genetic code. In DNA the four nucleotide bases, adenine, guanine, cytosine and thymine (and in RNA the four bases, adenine, guanine, cytosine and uracil), constitute an alphabet of four letters. During translation the sequence of these bases is read in groups of three, each triplet of bases designating which of the twenty or more available amino acids is incorporated into the growing polypeptide chain.

Genome. A collective term for all the genes contained in a single (haploid) set of chromosomes.

Globular proteins. Proteins in which the polypeptide chains are folded into compact globular shapes.

Examples include most enzymes, antibodies and many hormones.

Glycolysis. The production of chemical energy by the anaerobic breakdown of glucose to lactic acid.

Glycoproteins. Large protein/carbohydrate molecules in which the sugar chains are covalently linked to the polypeptide 'backbone'.

Gonadotrophins. Glycoprotein hormones secreted by the pituitary gland. They control the activity of the gonads and stimulate the maturation of ova and sperm.

Haemoglobin. Major constituent of the red blood cell cytoplasmic matrix. Mol. wt. 65 000; consists of four polypeptide (globin) chains and an iron-containing haem group.

Haploid. Possession of a single set of chromosomes.

Homogenate. The mixture of cell organelles and sealed membrane fragments produced by the physical disruption of the cell.

Hormones. Molecules responsible for carrying messages between separate cells. Hormones are secreted by endocrine cells and are carried to their 'target cells' by the blood and other body fluids.

Hydrated. Containing chemically combined water. Opposite of anhydrous.

Hydrophilic/hydrophobic. Water attracting/water repelling.

Immunoglobulins. See *Antibodies*.

Insulin. A polypeptide hormone. (mol. wt. 12 000) secreted into the blood by the beta cells of the endocrine pancreas. Its key target tissues are the liver, muscle and fat cells, where it promotes the synthesis of (and/or inhibits the degradation of) glycogen, protein and lipid.
It is an absolute requirement for normal growth in man. Insufficient insulin causes diabetes mellitus, the most characteristic feature of which is hyperglycaemia — a raised blood glucose level.

Intercalated discs. Specialized plasma membrane junctional areas that connect cardiac muscle cells in series. They include desmosomal and gap junctional components.

In vitro/In vivo. From the Latin, in vitro ('in glass'), in vivo ('in life'). In this book, 'in vitro' includes all preparations (including cell culture) maintained outside the body. Elsewhere its use may be more restricted and reserved for non-cellular (i.e. broken cell) preparations.

Ion. Electrically charged atom. Cations are positively charged ions that have fewer electrons than is necessary for the atom to be electrically neutral; negatively charged ions (anions) have more. The proton (the hydrogen atom without its circumnuclear electron) is thus a hydrogen ion.

Isotropic/Anisotropic. Substances or components that display equal light properties in all directions are said to be 'isotropic'. Those that display different light properties in different directions are 'anisotropic' ('birefringent'). Examined in a polarizing microscope, striated muscles display alternating isotropic (I) and anisotropic (A) bands.

Karyotype. The number and morphological characteristics of the chromosomes of a cell. In the diploid somatic cells of an individual the karyotype is constant.

Lectin. From the Latin 'legere', 'to choose', 'to select'. In cell biology it is used in reference to a group of protein and glycoprotein molecules that bind specifically to cell surface sugars.

Leucocytes. Nucleated cells of the peripheral blood. They include granular forms (neutrophils, eosinophils and basophils) and the 'non-granular' lymphocytes and monocytes.

Lipids. A heterogeneous class of structural types most easily defined as those molecules in biological material which are extractable by non-polar solvents (i.e. 'fat solvents' like ether and chloroform).

Lymphocyte. Free cell, 8 to 16 μm in diameter, with a large heterochromatic nucleus surrounded by pale-staining 'undifferentiated' cytoplasm. The lymphocyte is a major participant in all kinds of immune response and a major constituent of the blood and lymphatic tissues (such as lymph node, thymus, spleen and tonsil). Functionally, there are several different kinds of lymphocyte. Amongst the most important are the 'B' lymphocytes (responsible for generating the humoral or antibody-related response) and the thymus-derived 'T' lymphocytes (important in cell-mediated responses like graft rejection).

Macromolecules and macromolecular assemblies. Macromolecules include nucleic acids, proteins, polysaccharides and lipids; large molecules with molecular weights of between 10^3 and 10^9. They may be grouped to form 'macromolecular assemblies' — independent functional units such as ribosomes and chromosomes.

Macrophages. Phagocytic, often freely mobile cells. Largely derived from the monocytes of the blood and part of the 'mononuclear phagocyte' system.

Meiosis. The process of division that reduces a diploid chromosome number to the haploid amount present in germ cells (ova and sperm).

Mitosis. The process of division that maintains the same (diploid) chromosome number in the progeny as in the parent cell.

Molecular weight (mol. wt.). Sum total of all the constituent atoms in a molecule. Usually expressed in Daltons.

Monolayer. A layer of cells one cell thick; in culture many kinds of cell grow as a monolayer.

Monomer. Repeated, identical subunits that make up a polymer. Unlike monomers, protomers and subunits need not be identical with each other.

Mordants. Salts of iron and aluminium that react with certain stains (such as haematein) to form a positively-charged complex known as a 'lake'. The advantages of a lake are that it penetrates better, binds more avidly and when bound to a tissue component is more insoluble than the uncomplexed stain.

N terminus. See *Peptide bond*.

Negative staining. Method of introducing contrast into particles (or fragments) for electron microscopy. The particles are dispersed over the surface of a supporting film and then impregnated with a thin film of heavy metal salt (such as uranyl acetate). Unimpregnated regions stand out in negative relief against the electron-opaque background provided by the stain.

Neurones. Most neurones have a cell body (containing the nucleus) which gives rise to a single effector axon and a variable number of dendrites. Axons terminate as synapses; conduction of an impulse along the axon induces the release of a neurotransmitter from the synapse.

Neurotransmitters. Chemical messenger molecules released by neurones and which act (i.e. bind to receptors on their target cell) within the vicinity of their site of release. They include acetylcholine and noradrenaline. Messenger molecules released by neurones and which have a distant site of action are called 'hormones'.

Nucleosides/Nucleotides. Nucleosides are molecules in which ribose or deoxyribose sugars are linked to pyrimidines (such as thymine, uracil and cytosine) or purines (such as adenine and guanine). Nucleotides are phosphoric esters of nucleosides in which the phosphate is esterified to the sugar. They include not only the nucleotides, which are constituents of nucleic acids, but also those like adenosine monophosphate, which can be further phosphorylated to the high-energy compounds adenosine diphosphate and adenosine triphosphate.

Oestrogens. Female sex steroid hormones of which the most important are oestradiol-17β and oestrone.

Opsonin. From the Greek 'opsonein', 'to prepare food for'. Antibody which coats invading bacteria and increases the probability that they will be phagocytosed.

Organelle. Membrane-limited compartments responsible for specific cellular functions. Interdependent with the cell as a whole but often capable of self-replication.

Osteoclasts. Large multinucleate cells that work in conjunction with osteoblasts to sculpture bone. Osteoblasts form bone matrix, osteoclasts degrade and remove it.

Ovalbumin. Protein synthesized in large amounts by hen oviduct cells; a major component of egg white.

Oxidation. Addition of oxygen or removal of hydrogen. Includes also any reaction in which an atom loses electrons, as, for example, when a ferrous ion (Fe^{++}) changes to a ferric ion (Fe^{+++}).

β **Oxidation** of fatty acids. Fatty acyl-CoA thioesters entering the mitochondrion from the cytoplasm are broken down by the successive oxidative removal of acetyl-CoA units. The electrons produced during this process enter the electron transport chain while the acetyl-CoA's enter the TCA cycle.

Peptide bond. In proteins, the carboxyl group (COOH) of one amino acid is connected (with the loss of a molecule of water) to the amino group (NH_2) of the next. Proteins therefore have a free amino group at one end (often referred to as the 'N terminal' end) and a free carboxyl group (often referred to as the 'C terminal') at the other.

Phospholipid. Like most lipids, phospholipids are built upon the 3-carbon unit of glycerol. Two of these carbons are associated with long fatty acid chains (and these chains are usually different from each other) while the third carbon is associated with a polar group that contains phosphate. The polar group carries (or can accept) an electrical charge and is therefore hydrophilic.

Phosphorylation. A reaction mediated by a group of enzymes called 'protein kinases', which transfer a phosphate group from a 'high-energy' donor (usually ATP) to an acceptor protein. When phosphorylated, the functional properties of the acceptor protein are altered. The reaction can be reversed by a group of enzymes known as 'phosphatases'.

Plaque. A patch or spot on a surface.

Plasma cell. A member of the immune system, responsible for synthesizing and secreting large amounts of antibody. The antibody synthesized by each plasma cell is specific for a particular antigen.

Platelets. Enucleate discoid constituents of peripheral blood. Injury to the wall of a blood vessel stimulates platelets to secrete a variety of active agents that together cause blood clotting (i.e. coagulation of the plasma and cell adhesion). Stimulated platelets lose their discoid shape and become contractile.

Polar molecules. Molecules which have within them an uneven distribution of electrical charge possess an 'electrical dipole moment' and are called polar molecules. Because their charged groups are able to form hydrogen bonds with water molecules, polar molecules are hydrophilic. Non-polar molecules (or non-polar regions of molecules) are hydrophobic.

Polarity. With a preferred direction.

Polymer. A chain of repeated units. In biological molecules they are usually covalently linked.

Polymerases. Enzymes that promote the synthesis of polymers. Used in this book with reference to nucleic acid synthesis, where nucleotides and deoxynucleotides are assembled into RNA and DNA respectively.

Polynucleotide chain. Linear sequence of nucleotides linked by phosphate groups.

Prokaryotes. Lower forms of life like bacteria in which the genetic material lies free within the cell (i.e. there is no nucleus).

Proteases. Class of enzymes which degrade protein (proteolysis). Exemplified by trypsin, a pancreatic protease which cleaves polypeptide chains whenever it has access to adjacent lysine and arginine residues (see *Cathepsins*).

Proteins. Linear polymers of amino acids that may be twisted, pleated or folded. The amino acid chains do not themselves branch although, as in glycoproteins, they may give rise to side chains of sugar residues.

Protomer. See *Monomer*.

Puromycin. Inhibitor of protein synthesis. In translation it is incorporated in place of an amino acyl—tRNA and terminates the growing polypeptide chain.

Radioactive isotopes. Isotopes are atoms of the same element which (although otherwise identical) contain different numbers of neutrons in their nuclei. In radioactive isotopes the atomic nucleus is unstable and disintegrates spontaneously. Disintegration (the conversion of a neutron) is accompanied by the emission of either alpha or beta particles and/or gamma rays. ^{3}H (tritium) is a beta-particle emitter; ^{125}I emits gamma rays and beta particles.

S (Svedberg) unit. The unit of sedimentation. It is proportional to the rate of sedimentation of a particle (molecule) in a specified centrifugal field. The 'S' value is determined partly by the shape of a molecule, but it nevertheless gives a useful indication of molecular size.

Schwann cells. In the peripheral nervous system, axons are intimately associated with attendant Schwann cells. Along the axons of 'myelinated' neurones, regularly arranged Schwann cells provide a linear series of insulating sheaths. Nodes of Ranvier mark the uninsulated intervals between each of these consecutive sheaths.

Skeletal (voluntary) muscle. Voluntary muscles composed of long (up to 40 μm) multinucleate fibres. Each fibre tapers at the ends and inserts directly into the anchoring connective tissues.

Both skeletal and cardiac muscles fibres are striated and in both of them the contractile unit is the sarcomere.

Smooth muscle. Involuntary muscle composed of bundles or sheets of individual, spindle-shaped cells. As a sleeve-like layer surrounding blood vessels, smooth muscle controls the size of the lumen. Around the tubular pathways of the intestines and reproductive tracts, smooth muscles may give rise to 'peristalsis', a synchronized flow of contraction that moves the luminal content along.

Somatic cells. All the cells of the body other than germ cells.

Stem cells. Undifferentiated cells of adult tissues retaining the capacity for continued mitosis. They probably occur in all tissues where the differentiated cells need to be continuously replaced.

Steroids. A large class of molecules with diverse biological activities. They are derivatives of the fused ring system, perhydrocyclopentenophenanthrene:

The most abundant steroid in animal tissues is cholesterol; others include the hormonal steroids of the adrenal cortex (e.g. cortisol and 17 β-hydrocorticosterone) and the sex steroids (e.g. oestrogen and oestradiol-17β, secreted by the ovary; and progesterone, secreted by the corpus luteum).

Subunit. See *Monomer*.

Supernatant. Those components that remain unsedimented following each step in differential centrifugation.

Target cell. See *Hormones*.

Template. The mould or framework of one molecular component used to determine or select the form of another.

Units.

1 metre (1 m) = 1000 millimetres (mm)
1 millimetre (1 mm) = 1000 micrometres (μm)
1 micrometre (1 μm) = 1000 nanometres (nm)

(1 nm = 10 Ångströms, Å)

Virus. Disease-producing particle dependent upon other organisms for multiplication. Consists of either DNA or RNA plus protein.

Vitamin. Essential organic molecule that must be provided in diet.

Zymogen. Inactive enzyme precursor.

BIBLIOGRAPHY

BOOKS AND ARTICLES OF GENERAL INTEREST

Lehninger, A. L. (1975) *Biochemistry*. 2nd Edn. 1044 pp. New York: Worth. *First-rate text with a strong emphasis on cellular aspects.*

Watson, J. D. (1975) *Molecular Biology of the Gene*. 3rd Edn. 739 pp. Menlo Park, California: W. A. Benjamin. *Entertaining, well-illustrated book now extended to include a good deal of cell biology.*

Streyer, H. (1975) *Biochemistry*. San Francisco: Freeman. *Well-illustrated general text.*

Loewy, A. G. & Siekevitz, P. (1969) *Cell Structure and Function*. 2nd Edn. 512 pp. New York: Holt, Rinehart & Winston. *Good basic cell biology text with a biochemical bias.*

Bloom, W. & Fawcett, D. W. (1975) *A Textbook of Histology*. 10th Edn. Philadelphia: W. B. Saunders. 1033 pp. *Superb, all-round histology text copiously illustrated.*

DeRobertis, E. D. P., Saez, F. A. & DeRobertis, E. M. F. (1975) *Cell Biology*. 6th Edn. 618 pp. Philadelphia: W. B. Saunders.

Novikoff, A. B. & Holtzmann, E. (1976) *Cells and Organelles*. 2nd Edn. New York: Holt, Rinehart & Winston. 400 pp. *Well-written stimulating text with the emphasis on structure.*

Ganong, W. F. (1977) *Review of Medical Physiology*. 8th Edn. Los Altos, California: Large Medical Publications. 587 pp. *Brief, well-presented up-to-date outline.*

Davson, H. (1970) *A Textbook of General Physiology*. Vols. 1 and 2. 4th Edn. Edinburgh: Churchill Livingstone. *Basic, comprehensive but slightly out of date.*

Goodenough, U. & Levine, R. P. (1978) *Genetics*. 2nd Edn. 864 pp. New York: Holt, Rinehart & Winston. *Comprehensive, up-to-date text.*

White, M. J. D. (1972) *The Chromosomes*. 6th Edn. New York: John Wiley. 108 pp. *Good introduction to the structure and function of chromosomes.*

Davis, B. D., Dulbecco, R., Eisen, H. N., Ginsberg, H. SS, Wood, W. B. Jr and McCarty, M. (1973) *Microbiology*. 2nd Edn. 1562 pp. London: Harper & Row. *Excellent but massive text in medical microbiology. Basic immunology particularly well done.*

Roitt, I. (1974) *Essential Immunology*. 2nd Edn. Oxford: Blackwell. 220 pp. *Well-written general introduction to immunology.*

Anfinsen, C. B. (1973) Principles that govern the folding of protein chains. *Science*, **181**, 223–230. *Showing that the amino acid sequence determines the three-dimensional conformation of protein molecules and thus their biological function.*

Spiro, R. G. (1969) Glycoproteins. Their biochemistry, biology and role in human disease. *New England Journal of Medicine*, **281**, 991–1001, 1043–1056.

Kushner, D. (1969) Self assembly of biological structures. *Bacteriological Reviews*, **33**, 302. *Provides a useful insight into macromolecular organization.*

THE CELL THEORY

Baker, J. R. The cell theory: a restatement, history and critique. *Quarterly Journal of Microscopical Science*, **89**, 103–125 (1948); **90**, 87–108 (1949); **93**, 157–189 (1952); **94**, 407–440 (1953); **96**, 449–481 (1955). *Authoritative account of cytology in the nineteenth and early twentieth centuries.*

Hughes, A. (1959) *A History of Cytology*. New York: Abelard-Schuman.

Flemming, W. (1965) Contributions to the knowledge of the cell and its processes. Translated from the original article (1880) and reprinted in *Journal of Cell Biology*, **25**, 1–69. *The first comprehensive treatment of mitosis.*

Ramon Y Cajal, S. (1967) The structure and connections of neurons. Nobel lectures, Physiology or Medicine 1901–1921. Amsterdam: Elsevier.

Golgi, C. (1967) The neuron doctrine – theory and facts. Nobel lectures, Physiology or Medicine 1901–1921. Amsterdam: Elsevier. *This and the two preceding papers exemplify the approach and level of analysis achieved in the classical studies of the late nineteenth and early twentieth centuries.*

Porter, K. R. (1974) The 1974 Nobel Prize for Physiology or Medicine. *Science*, **186**, 516–520.

Palade, G. E. (1977) Keith Roberts Porter and the development of contemporary cell biology. *Journal of Cell Biology*, **75**, D4–D19. *This and the preceding paper describe the growth and impact of biological electron microscopy.*

MICROSCOPY AND RELATED TECHNIQUES

James, J. (1976) *Light Microscope Techniques in Biology and Medicine*. The Netherlands: Martinus Nijhoff Medical Division. *All aspects of light microscopy well covered.*

Baker, J. R. (1970) *Principles of Biological Microtechnique*. London: Methuen. 334 pp. *Histological stains, their history, chemistry and use.*

Wischnitzer, S. (1962) *Introduction to Electron Microscopy*. Oxford: Pergamon Press. 132 pp. *Clear, easily understood account.*

Glauert, A. M. (Ed.) *Practical Methods in Electron Microscopy*. Vol. 2: Principles and practice of electron microscope operation (1974). Vol. 3, Pt. I: Fixation, dehydration and embedding (1974). Vol. 3, Pt. II: Ultracryotomy (1974). Vol. 5: Staining methods for sectioned material (1977). Amsterdam: North Holland. *Useful continuing series for those wishing to use the technique.*

Pearse, A. G. E. (1972) *Histochemistry, Theoretical and Applied*. Vols. I and II. Edinburgh: Churchill Livingstone. *Encyclopaedic text with detailed recipes in the appendices.*

Rogers, A. W. (1967) *Techniques of Autoradiography*. Amsterdam: Elsevier. *Dated but full account.*

Pollack, R. (Ed.) (1975) *Readings in Mammalian Cell Culture*. 2nd Edn. New York: Cold Spring Harbor Laboratory. *Collection of articles illustrating the usefulness of cultured cells for the study of cell proliferation.*

Gospodarowicz, D. & Moran, J. S. (1976) Growth factors in mammalian culture. *Annual Review of Biochemistry*, **45**, 531–558. *An account which illustrates the potency and specificity of these endogenous factors.*

Birnie, G. D. (Ed.) (1972) *Subcellular Components: Preparation and Fractionation*. 2nd Edn. Baltimore: University Park Press.

THE PLASMA MEMBRANE

Historical

Branton, D. & Park, R. B. (1968) *Papers on Biological Membrane Structure*. Boston: Little, Brown. *The classical papers collected.*

Frye, L. D. & Edidin, M. (1970) The rapid intermixing of cell surface antigens after formation of mouse–human heterokaryons. *Journal of Cell Science*, **7**, 319–335. *The first definitive demonstration of membrane fluidity.*

Singer, S. J. & Nicolson, G. C. (1972) The fluid mosaic model of the structure of cell membranes. *Science*, **175**, 720.

Abercrombie, M. & Heaysman, J. E. M. (1954) Observations on the social behaviour of cells in tissue culture. II. 'Monolayering' of fibroblasts. *Experimental Cell Research*, **6**, 293–310. *The first account of contact inhibition.*

Hodgkin, A. L. (1964) The ionic basis of nervous conduction. Nobel lecture, 1963. *Science*, **145**, 1148–1153.

Books and articles providing useful background information

Nystrom, R. A. (1973) *Membrane Physiology*. New Jersey: Prentice Hall.

Katz, B. (1966) *Nerve, Muscle, Synapse*. New York: McGraw-Hill. 193 pp. *Clear, very readable account.*

Sutherland, E. W. (1972) Studies on the mechanism of hormone action. *Science*, **177**, 401–407.

Stein, W. D. (1967) *Movement of Molecules Across Membranes*. New York: Academic Press.

Sharon, N. (1977) Lectins. *Scientific American*, **236**, 108–119. *Clear, general overview.*

Current reviews

Weissmann, G. & Claiborne, R. (Ed.) (1975) *Cell Membranes; Biochemistry, Cell Biology and Pathology*. New York: Hospital Practice. 283 pp. *Collection of general articles by many of the leading workers in the field. Strong medical bias. Excellent illustrations.*

Cautrecasas, P. (1974) Membrane receptors. *Annual Review of Biochemistry*, **43**, 169–215. *Full coverage of receptors for peptide and polypeptide hormones.*

Goldstein, J. L. & Brown, M. S. (1977) The low density lipoprotein pathway and its relation to atherosclerosis. *Annual Review of Biochemistry*, **46**, 897–930. *Concise account by the major contributor in this area.*

Raff, M. C. (1976) Cell surface immunology. *Scientific American*, **234**, No. 5, 30–39. *Lively, beautifully illustrated account of the lymphocyte plasma membrane.*

Nicolson, G. L., Giotta, G., Lotan, R., Neri, A. & Poste, G. (1977) Modifications in transformed and malignant tumour cells. *International Cell Biology 1976–1977*. pp. 138–148. (Ed.) Brinkley, B. R. & Porter, K. R. New York: Rockefeller University Press. *Full survey of a complex topic. A lucid distillation of the available information.*

Old, L. J. (1977) Cancer immunology. *Scientific American*, **236**, 63–73. *Tumour cell-surface antigens.*

Silverstein, S. C., Steinman, R. M. & Cohn, Z. A. (1977) Endocytosis. *Annual Review of Biochemistry*, **46**, 669–722. *Authoritative and comprehensive survey.*

Gilula, N. B. (1974) Junctions between cells. In *Cell Communication* (Ed.) Cox, R. P. New York: John Wiley. pp. 1–29. *Brief account, emphasis on morphology and postulated function.*

Articles

Revel, J. P., Hoch, P. & Ho, D. (1974) Adhesion of

culture cells to their substratum. *Experimental Cell Research*, **84**, 207–218.

Rees, D. A., Lloyd, C. W. & Thom, D. (1977) Control of grip and stick in cell adhesion. *Nature*, **267**, 124–128.

Deguchi, N., Jørgensen, P. L. & Maunsbach, A. B. (1977) Ultrastructure of the sodium pump. *Journal of Cell Biology*, **75**, 619–634.

THE NUCLEUS

Historical

Watson, J. D. & Crick, F. H. C. (1953) Molecular structure of nucleic acids: a structure for deoxypentose nucleic acids. *Nature*, **171**, 737. *The original description of the DNA helix.*

Mirsky, A. E. (1968) The discovery of DNA. *Scientific American*, **218**, 78–88. *Historical perspective.*

Jacob, F. & Monod, J. (1961) Genetic regulatory mechanisms in the synthesis of proteins. *Journal of Molecular Biology*, **3**, 318–356. *The link between messenger RNA and protein synthesis.*

Palade, G. E. (1955) A small particulate component of the cytoplasm. *Journal of Biophysical and Biochemical Cytology*, **1**, 59–68. *Identification of the ribosome in the electron microscope.*

Books and reviews

Watson, J. D. (1975) *The Molecular Biology of the Gene*. 3rd Edn. Menlo Park, California: W. Benjamin, 739 pp.

Haynes, R. H. & Hanawalt, P. C. (1968) *The Molecular Basis of Life*. San Francisco: Freeman. *Collection of* Scientific American *articles providing a good background to classic studies in the molecular biology of nucleic acids. Articles by Crick, Holley, Taylor, Rich, Nirenberg and Clark and Marker especially useful.*

Kornberg, A. (1974) *DNA Synthesis*. San Francisco: Freeman. *An advanced, encyclopaedic text.*

Structure and Function of Chromatin (1975). CIBA Foundation Symposium 28 (New Series). Amsterdam: Elsevier. *Collection of advanced articles by the leaders in the field.*

Kornberg, R. D. (1977) Structure of chromatin. *Annual Review of Biochemistry*, **46**, 931–954. *Recent but primarily structural treatment.*

Mazia, D. (1974) The cell cycle. *Scientific American*, **230**, 55–64. *Excellent well-written and well-illustrated account.*

Padilla, G. M., Cameron, I. L. & Zimmerman, A. (1974) *Cell Cycle Controls*. 370 pp. New York: Academic Press. *Collected advanced articles, very informative.*

Clarkson, B. & Baserga, R. (Ed.) (1974) *Control of Proliferation in Animal Cells*. New York: Cold Spring Harbor Press. *Encyclopaedic collection of research papers.*

Lewin, B. M. (1970) *The Molecular Basis of Gene Expression*. New York: Wiley–Interscience.

Chauberlin, M. J. (1974) The selectivity of transcription. *Annual Review of Biochemistry*, **43**, 721–775.

Davidson, E. H. & Britten, R. J. (1973) Organization, transcription and regulation in the animal genome. *Quarterly Review of Biology*, **48**, 565–613. *Regulation by nucleic acids – an hypothesis.*

Stein, G. S., Spelsberg, T. C. & Kleinsmith, L. J. (1974) Non histone chromosomal proteins and gene regulation. *Science*, **183**, 817–823.

Tomkins, G. M. (1974) Regulation of gene expression in mammalian cells. Harvey Lectures, **68**, 37–66. *The role of steroids.*

O'Malley, B. W. & Schrader, W. T. (1976) The receptors of steroid hormones. *Scientific American*, **234**, No. 2, 32–43. *The best-understood system of control over eukaryotic gene expression.*

Maniatis, T. & Patashine, M. (1976) A DNA operator-repressor. *Scientific American*, **234**, 64–76. *A review that shows the level of analysis now achieved.*

Nomura, M., Tissieres, A. & Lengyel, P. (Eds.) (1974) *Ribosomes*. New York: Cold Spring Harbor Laboratory of Quantitative Biology. *Collection of advanced reviews.*

Haselkorn, R. & Rothman-Denes, L. B. (1973) Protein synthesis. *Annual Review of Biochemistry*, **42**, 397–438. *Molecular basis of translation, based mostly on prokaryotic studies.*

Blobel, G. (1977) Synthesis and segregation of secretory proteins: the signal hypothesis. pp. 318–325. *International Cell Biology 1976–1977*. (Ed.) Brinkley, B. R. & Porter, K. R. New York: Rockefeller University Press. *Brief, clear account of the signal hypothesis and its implications.*

INTRACELLULAR COMPARTMENTS AND CYTOPLASMIC ORGANELLES

Historical

Warburg, O. H. (1965) The oxygen-transferring ferment of respiration. Nobel lectures, Physiology or Medicine, 1922–1941, p. 254. Amsterdam: Elsevier.

Krebs, H. A. (1964) The citric acid cycle. Nobel lectures, Physiology or Medicine 1942–1962, p. 399. Amsterdam: Elsevier.

Lipmann, F. (1941) Metabolic regulation and utilization of phosphate bond energy. *Advances in Enzymology*, **1**, 99. *The above three accounts encapsulate each author's contribution to our understanding of how cells obtain and handle energy.*

Palade, G. E. & Porter, K. R. (1954) Studies on the endoplasmic reticulum. I. Its identification in cells *in situ. Journal of Experimental Medicine*, **100**, 641–656. *Not the first citing but the first full appreciation of the form and extent of the rough endoplasmic reticulum.*

Porter, K. R. & Palade, G. E. (1957) Studies on the endoplasmic reticulum. III. Its form and distribution in striated muscle cells. *Journal of Biophysical and Biochemical Cytology*, **3**, 269.

The first definitive account of the sarcoplasmic reticulum.

de Duve, C. (1969) The lysosome in retrospect. In *Lysosomes in Biology and Pathology*, Vol. 1. (Ed.) Dingle, J. T. & Fell, H. B. Amsterdam: North Holland. *An entertaining account of how the lysosome was discovered.*

Mitchell, P. (1961) Coupling of phosphorylation to electron and hydrogen transfer by a chemiosmotic type of mechanism. *Nature*, **191**, 144. *The original proposal; see the article below by Boyer et al (1977) for its current status.*

Reviews

Palade, G. E. (1975) Intracellular aspects of the process of protein secretion. Nobel lecture. *Science*, **189**, 347–358. *Masterly contemporary account of the secretory process.*

Leblond, C. P. (1977) Role of the Golgi apparatus in terminal glycosylation. pp. 326–336. *International Cell Biology 1976–1977*. (Ed.) Brinkley, B. R. & Porter, K. R. New York: Rockefeller University Press. *Brief, authoritative description of the final stages in glycoprotein manufacture.*

Beams, H. W. & Kessel, R. G. (1968) The Golgi apparatus: structure and function. *International Review of Cytology*, **23**, 209–276. *Full coverage of light and electron microscope studies.*

Steiner, D. F., Kemmler, W., Tager, H. S. & Peterson, J. D. (1974) Proteolytic processing in the biosynthesis of insulin and other proteins. *Federation Proceedings*, **33**, 2105–2115. *Brief review building on the classical proinsulin study.*

Dean, R. T. & Barrett, A. J. (1976) Lysosomes. *Essays in Biochemistry*, **12**, 1–40.

Neufeld, E. F., Lim, T. & Shapiro, L. (1975) Inherited disorders of lysosomal metabolism. *Annual Review of Biochemistry*, **44**, 357–371. *An important medical aspect of lysosomal activity defined.*

Lehninger, A. L. (1965) *The Mitochondrion: Molecular Basis of Structure and Function*. Menlo Park, California: W. A. Benjamin.

de Duve, C. (1969) The peroxisome: a new cytoplasmic organelle. *Proceedings of the Royal Society B*, **173**, 71–83.

Recent articles

MacLennan, D. H. & Holland, P. (1975) Calcium transport in the sarcoplasmic reticulum. *Annual Review of Biophysics and Bioengineering*, **4**, 377–404.

Zubrzycka, E. & MacLennan, D. H. (1977) Assembly of the sarcoplasmic reticulum. *Journal of Biological Chemistry*, **251**, 7733–7738.

Boyer, P. D., Chance, B., Ernster, L., Mitchell, P., Racker, E. & Slater, E. C. (1977) Oxidative phosphorylation and photophosphorylation. *Annual Review of Biochemistry*, **46**, 955–1025. *Collective analysis of current concepts.*

THE CYTOPLASMIC MATRIX

Historical

Needham, D. M. (1971) *Machina Carnis. The Biochemistry of Muscular Contraction in Historical Development*. London: Cambridge University Press.

Huxley, H. E. & Hanson, J. (1954) Changes in the cross-striations of muscle during contraction and stretch and their structural interpretation. *Nature*, **173**, 973–976. *The sliding filament hypothesis proposed.*

Reviews

Squire, J. M. (1975) Muscle filament structure and muscle contraction. *Annual Review of Biophysics and Bioengineering*, **4**, 137–163. *Clear account of force generation.*

Weber, A. & Murray, J. M. (1973) Molecular control mechanism in muscle contraction. *Physiological Reviews*, **53**, 412–673. *Full discussion of control in striated muscle.*

Somlyo, A. P. & Somlyo, A. V. (1976) Ultrastructural aspects of activation and contraction of vascular smooth muscle. *Federation Proceedings*, **35**, 1288–1293. *Concise account of smooth muscle structure.*

Pollard, T. D. (1977) Cytoplasmic contractile proteins. *International Cell Biology 1976–1977*. (Ed.) Brinkley, B. R. & Porter, K. R. New York: Rockefeller University Press. *Brief account with the emphasis on the mechanisms of cell movement.*

Wessells, N. K. (1974) How living cells change shape. *Scientific American*, **225**, 77. *The role of microfilaments and microtubules.*

Trinkaus, J. P. (1976) On the mechanisms of metazoan cell movements. pp. 225–329. In *The Cell Surface Reviews*, Vol. 1. (Ed.) Poste, G. & Nicolson, G. L. Oxford: North Holland. *Clear, comprehensive review: an excellent background to a complex area.*

Olmsted, J. B. & Borisy, G. G. (1973) Microtubules. *Annual Review of Biochemistry*, **42**, 507–540. *Not the most recent but still one of the best reviews for all-round coverage.*

Mohri, H. (1976) The functions of tubulin in motile systems. *Biochimica et Biophysica Acta*, **456**, 85–127. *Strong biochemical emphasis.*

Satir, P. (1974) How cilia move. *Scientific American*, **231**, 45–52. *Detailed account of morphological studies.*

Recent articles

Mooseker, M. S. & Tilney, L. G. (1975) Organization of an actin filament–membrane complex. *Journal of Cell Biology*, **67**, 725–743. *The dissection of the brush border contractile system.*

Lazarides, E. & Hubbard, B. D. (1976) Immunological characterization of the subunit of the 100 Å filaments from muscle cells. *Proceedings of the National Academy of Sciences*, **73**, 4344–4348. *Distribution of desmin.*

Heggeness, M. H., Wang, K. & Singer, S. J. (1977) Intracellular distributions of mechanochemical proteins in cultured fibroblasts. *Proceedings of the National Academy of Sciences*, **74**, 3883–3887. *Distribution of actin, myosin and filamin in non-muscle cells.*

Craig, R. & Megerman, J. (1977) Assembly of smooth muscle myosin into side-polar filaments. *Journal of Cell Biology*, **75**, 990–996. *An important difference in the arrangement of myosin molecules in smooth muscle.*

Small, J. Y. & Sobieszek, A. (1977) Ca regulation of mammalian smooth muscle actomyosin via a kinase-phosphate-dependent phosphorylation and depolymerization of the 20 000-Mr light chain of myosin. *European Journal of Biochemistry*, **76**, 521–530.

INDEX